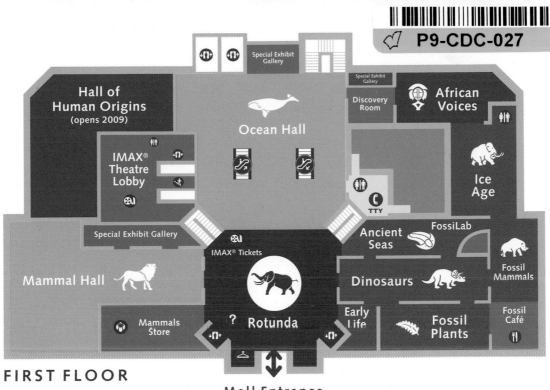

Hall of Human Origins (opens 2009)

IMAX® Theatre Lobby

Ocean Hall

Special Exhibit Gallery

Discovery Room

African Voices

Ice Age

Special Exhibit Gallery

Mammal Hall

Ancient Seas

FossiLab

Dinosaurs

Fossil Mammals

IMAX® Tickets

? Rotunda

Mammals Store

Early Life

Fossil Plants

Fossil Café

FIRST FLOOR

Mall Entrance

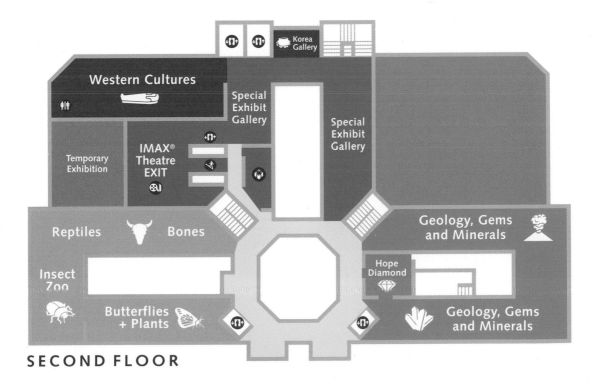

Western Cultures

Korea Gallery

Special Exhibit Gallery

Special Exhibit Gallery

Temporary Exhibition

IMAX® Theatre EXIT

Reptiles **Bones**

Geology, Gems and Minerals

Insect Zoo

Hope Diamond

Butterflies + Plants

Geology, Gems and Minerals

SECOND FLOOR

OFFICIAL GUIDE TO THE SMITHSONIAN
National Museum of Natural History

REVISED EDITION

OFFICIAL GUIDE TO THE SMITHSONIAN

National Museum of Natural History

REVISED EDITION

SMITHSONIAN BOOKS • WASHINGTON, DC

Library of Congress Cataloging-in-Publication Data
Official guide to the Smithsonian National Museum of Natural History.
    p.   cm.
   ISBN 1-58834-109-7 (alk. paper)
    1. National Museum of Natural History (U.S.)—Guidebooks. 2. Natural history museums—Washington (D.C.)—Guidebooks. I. Smithsonian Institution.

QH70.U62W275 2004
508'.074'753—dc22   2003057377

Printed in China, not at government expense

09 10 11 12 13   5 4 3 2 1

GUIDEBOOK STAFF
Writers: Sharon L. Barry, Robin A. Faitoute, Sarah Grusin, and Elizabeth Jones
Project coordinator: Lorena Selim
Image manager: Michael Atwood Mason

SMITHSONIAN BOOKS
Executive editor: Caroline Newman
Production editor: Robert A. Poarch, Christina Wiginton
Editorial proofreader: Laura A. Starrett, Lise Sajewski
Designer: Original design by Brian Barth; Updated design by Jody Billert / Design Literate, Inc.

Overleaf: View of the elephant diorama in the Kenneth E. Behring Family Rotunda.

Opposite: Peresphinctes, fossil ammonites, from the Middle Jurassic period

ACKNOWLEDGMENTS
We would like to thank the many individuals at the National Museum of Natural History and other Smithsonian divisions who reviewed the manuscript, assisted in photo research, or provided images: Mary Jo Arnoldi, Carole Baldwin, Michael Brett-Surman, David Burgevin, William DiMichele, Richard Efthim, Lynn Ellington, Robert Emry, Douglas Erwin, Nathan Erwin, Terry Erwin, Karen Fitzgerald, William Fitzhugh, Kathleen Gordon, Margery Gordon, Gary Graves, Candace Greene, Paula Healy, Pamela Henson, Robert Hoffmann, Brian Huber, Steven Jabo, David Johnson, Jill Johnson, Adrienne Kaeppler, Conrad Labandeira, Sally Love Connell, Michael Mason, Betty Meggers, Elizabeth Musteen, Vilma Ortiz-Sanchez, Lynne Parenti, Jeffrey Post, Richard Potts, Heather Rostker, Rusty Russell, Jerald Sachs, Carolyn Sadler, Kathy Sklar, Rebecca Snyder, Paul Taylor, Gus Van Beek, Michael Vecchione, Laura Viney, Scott Wing, Wendy Wiswall, and George Zug. We are especially grateful to the following Smithsonian photographers for taking so many new images for the book: Michael Barnes, Chip Clark, James DiLoreto, Donna Greene, Carl C. Hansen, Donald Hurlbert, and John Steiner. Finally, we would like to acknowledge Lorena Selim, assistant director for exhibits, for her support and encouragement.

CONTENTS

WELCOME TO THE MUSEUM

S tep inside the Smithsonian's National Museum of Natural History, our nation's premier natural history museum and a leader in the worldwide scientific community. Since it opened in 1910, more people have visited it than any other museum in the world. Nearly 7 million children and adults from every corner of the world walk through our doors each year. In addition, tens of millions visit us online at www.mnh.si.edu, where they find detailed information about our exhibits, events, collections, and educational programs.

The National Museum of Natural History is dedicated to understanding the natural world and our place in it, a mission carried out by our talented scientists, educators, and other professional staff. In this virtual age, the

An animatronic dinosaur is carried into the Museum for a temporary exhibit. There is always something new to see at the Museum.

Cristián K. Samper, director of the
National Museum of Natural History

Museum provides a special opportunity to see and interact with real things. You can examine stardust that is older than the Solar System, get nose-to-nose with some of the largest creatures to ever walk the Earth, and compare fossils of early humans that have come and gone over the past 6 million years. These treasures belong to the National Collections—an unsurpassed resource of more than 126 million natural and cultural objects that enables scientists and the public alike to better understand our world.

The Museum is also an important center for scientific discovery across many fields, and our research is increasingly reflected in the exhibitions. The diversity of living animals is on parade in our Biology halls, including the award-winning *Kenneth E. Behring Family Hall of Mammals* and the *Sant Ocean Hall*—one of the Museum's most ambitious projects yet. Unparalleled collections, state-of-the-art technology, and dramatic displays come together in the Ocean Hall to communicate a critical message—the ocean is a global ecosystem essential to all life on our planet. In the *Janet Annenberg Hooker Hall of Geology, Gems, and Minerals*—home of the Hope Diamond—you will see how tiny atoms came together to create the entire Solar System. You can walk through 3.5 billion years in the Fossils Halls and follow the evolution of life from single-celled creatures to enormous dinosaurs and woolly mammoths. Our Anthropology Halls present a stunning array of artifacts from cultures around the world.

In recent years, the Museum has undergone ambitious large-scale renovations, resulting in exciting new exhibition halls as well as several smaller galleries for temporary exhibitions and a variety of other educational experiences and events. The Johnson IMAX® Theater shows large-format films, some in 3-D. Staff at the information desks will be happy to tell you what special exhibitions, films, and events are available on the day you visit.

I hope that you enjoy your visit, and that you come back soon to see how the Museum continues to evolve.

Cristián Samper K.

Cristián K. Samper
Director, National Museum of Natural History

This 3.1 kg (6.8 lb.) aquamarine crystal and 1,000-carat gem are on display in the *Janet Annenberg Hooker Hall of Geology, Gems, and Minerals.*

NATIONAL MUSEUM OF NATURAL HISTORY: WHERE WONDERS NEVER CEASE

Almost a century ago, the brass doors of a young museum swung open on the National Mall. Since then, the National Museum of Natural History has evolved into an internationally recognized authority on the natural and cultural wonders of the world.

In 1910, the green-slate dome of this Museum rose above the tree line, forever changing the city's skyline and the nation's scholarly resources. The building was the third in the growing Smithsonian Institution. On the Mall, it was second in size only to the U.S. Capitol. With its growing collections, the new Museum established the Smithsonian as a major scientific and educational institution. Each passing year has built on that foundation.

The building's first name, the United States National Museum, reflected the broad scope of its

Today, the Museum's classical façade—with 13.8 m (45 ft.) tall Corinthian columns—is still an impressive sight from the National Mall.

Above: Today, the National Museum of Natural History covers nearly 1.6 ha (4 acres) on the Mall.

These zone-tailed hawk eggs were collected in 1872.

contents at that time. It was home to fine art and American history objects as well as cultural artifacts and natural history specimens. The Museum did not take its present name until 1968.

Today, the building stands in the company of nine other museums on the Mall. Through name changes, building additions, and reorganizations, the Museum has expanded and broadened its commitment to the pursuit and dispersion of knowledge.

THE ARCHITECTURE OF EXCELLENCE

When the National Museum of Natural History was completed in 1910, its Beaux Arts style made a strong architectural statement that validated the Smithsonian's newfound stature in the international community. Designed by Joseph C. Hornblower and J. Rush Marshall, this Museum was the first building on the Mall to realize the classical vision set out by federal city planners. The exterior columns and slate-covered dome create a memorable impression even today.

Inside, the four-story-high Rotunda provides an elegant and inspiring meeting place for visitor and scholar alike. Marble floors, limestone walls and piers, and architectural detail reinforce classical themes. Three floors of marble columns and natural light from fourth-floor windows draw the eye upward to the 38 m (125 ft.) high dome, an architectural symbol of heaven. It is covered with Guastavino tiles, a strong, lightweight tile seen in turn-of-the-twentieth-century domes around the world.

Three grand halls—each about 2,230 sq. m (24,000 sq. ft.)—radiate west, north, and east from the central Rotunda. Each features a central three-story atrium, graced with skylights. One hall was restored to its original grandeur for the opening of the *Kenneth E. Behring Family Hall of Mammals* in 2003, and another was completely renovated for the Museum's new *Sant Ocean Hall*.

The Constitution Avenue entrance serves as another welcoming space. White Vermont marble columns and walls, pink Tennessee marble pilasters, and plaster cornices and moldings recall some of the classical themes used in the Rotunda. Off the lobby, a central hall leads to many visitor services: the shops, cafeteria, and 565-seat Baird Auditorium.

In the early 1960s, two massive wings were added to the west and east sides of the Museum, providing six floors for collections storage, offices,

Marble columns and glowing chandeliers provide a distinguished backdrop for the displays in the Constitution Avenue Lobby.

Please
Don't Touch

From the Constitution Avenue Lobby, a hall leads past a *rai,* a symbol of status on the Micronesian island of Yap.

and research laboratories. Over the next three decades, the interior courtyards of the original building were filled with a café, shops, a three-story-high IMAX® theater, and a seven-floor office building. And as the Museum grows, its capacity to inspire curiosity and celebrate learning also continues to grow.

ONCE UPON A MUSEUM

If you came here between 1910 and 1920, you would have passed an outdoor market in front of the Museum. Once inside, you would have strolled past oil paintings, sculptures, a re-created farmyard, and some of the first mounted dinosaur bones. One popular 1913 diorama featured five lions collected by President Theodore Roosevelt during his famous hunting safaris. Although Roosevelt sent an estimated 6,000 specimens to the Smithsonian, the only one still on view is the white rhinoceros in the *Kenneth E. Behring Family Hall of Mammals.*

The legacy of other early exhibits is still in evidence today. The *Birds of the District of Columbia* survives, although many of the original bird specimens have been replaced. In *Reptiles: Masters of Land,* two of the earliest mounted dinosaurs in the world—a *Triceratops* and a *Stegosaurus*—have awed visitors since the turn of the century.

At one time, the Museum housed the history collections that are still a source of national pride. In 1913, visitors toured the Bliss Colonial Room to see period furniture and housewares. About forty years later, an entire seventeenth-century Massachusetts Bay Colony house was reconstructed on the second floor as part of a *Hall of Everyday Life in Early America.*

For the Museum's first few decades, a lively market existed on Constitution Avenue, then called B Street.

In 1957, to reflect its growing history collection, the National Museum changed its name to the Museum of Natural History and the Museum of History and Technology. Twelve years later, the history collections and staff moved into a new building next door, now called the National Museum of American History.

Part of the Smithsonian's art collection also occupied an uneasy fifty-year tenure here, sandwiched between two anthropology halls. Marble sculptures were sometimes grouped with mounted animal heads, and paintings were hung along hallways and in elevators. This Museum was once the home of several hundred George Catlin portraits of Native Americans. The art collection and staff finally earned their own home in the 1960s, relocating to the present-day National Portrait Gallery and the Smithsonian American Art Museum.

Above: National Gallery of Art staff enjoy a 1929 painting exhibit located on the first floor of the Museum.

In 1913, these five lions attracted crowds. President Theodore Roosevelt collected them on an African safari.

BUILDING THE COLLECTIONS

Today, the National Museum of Natural History houses more than 126 million specimens, artifacts, field notes, and photographs. Each year the collection swells by about 750,000 items. How did the Museum end up with such a staggering number of objects?

The National Collections began in 1846 with the formation of the Smithsonian Institution, which inherited boatloads and trainloads of artifacts and specimens collected by exploring expeditions of the 1830s. Thousands of these specimens and cultural objects became the nucleus of this Museum's collections. The National Herbarium, for example, includes nearly 50,000 plant specimens that were collected during the U.S. Exploring Expedition under the command of Lt. Charles Wilkes in 1838–42.

The fossil collection got its start during the 1880s, when the Museum sponsored research trips throughout the West. Some of the Museum's first dinosaur specimens were found in Wyoming rock formations and then transported by donkey and train back to Washington.

The Museum's world-renowned Burgess Shale Collection consists of 65,000 fossils that preserve amazingly fine details of animal life on Earth about 500 million years ago. The collection grew out of excavations by Charles D. Walcott, the fourth secretary of the Smithsonian, in western Canada during the early twentieth century.

Gifts from individuals have strengthened many parts of the collection. The Terry Anatomical Collection, received from Washington University's medical school, is named for anatomy professor Robert J. Terry, who began the collection in the early 1900s. Its 1,728 human skeletons have proven invaluable to anthropologists studying the effects of different diseases on bone.

This herbarium sheet, collected between 1838 and 1842, is among 4.5 million plant specimens housed in the Museum.

The Nelson Collection is the world's largest collection of Eskimo artifacts. During the 1880s, Army Signal Corps officer Edward W. Nelson was sent on a daring four-year-long expedition to Alaska, where he collected about 10,000 cultural objects.

The Mineral Science Collection has benefited from decades of generous contributions—including the 1958 donation of the Hope Diamond by Harry Winston and 16,000 mineral specimens from Washington A. Roebling, chief engineer of the Brooklyn Bridge.

Seemingly endless drawers of insects surround Museum entomologists. There are millions of insect specimens in our collection.

SPOTLIGHT SPENCER FULLERTON BAIRD:
A PASSIONATE COLLECTOR

As a teenager, Spencer Fullerton Baird roamed the Pennsylvania countryside collecting birds and other animals—an auspicious beginning for a man who would one day help build and shape the collections of this Museum.

In 1849, the twenty-seven-year-old Baird arrived in Washington, D.C., with his personal collection, which by then filled several boxcars. He came to work as an assistant secretary of the newly formed Smithsonian Institution. Baird became the institution's second secretary in 1878 and served until his death in 1887. As secretary, he ran the National Museum, then located in the Arts and Industries Building.

During Baird's nearly forty years of service, the collections grew so rapidly that the Smithsonian earned the nickname "the nation's attic." He helped train military personnel stationed in the West to collect specimens and artifacts, and he arranged for collecting expeditions to the Far North and South America.

Spencer Fullerton Baird's legacy survives in the auditorium that bears his name and the generations of distinguished scientists who have followed his example.

This oil painting by Henry Ulke (1887) of Spencer Fullerton Baird, the Smithsonian's second secretary, is part of the collection of the National Portrait Gallery.

OUTGROWING THE ATTIC

In 1978, a congressionally mandated collections inventory confirmed that the Museum was bursting at the seams. To provide room for the ever-growing collection, the Museum Support Center was built in Suitland, Maryland. Opened in 1983, this state-of-the-art facility provides 46,452 sq. m (500,000 sq. ft.) of storage space. It took twenty years for the Museum's anthropology department alone to plan, organize, and move 2 million cultural artifacts out there.

The Museum Support Center also contains specialized laboratories like the Laboratories of Analytical Biology, which store tens of thousands of DNA samples. In the future, tissue collections will serve as a kind of genetic baseline, and frozen sperm and embryos can be used in captive breeding programs to help endangered species.

The collections at the Museum Support Center and elsewhere in the Museum are available for study by scientists and scholars from around the world. Every year, the Museum loans about 200,000 cultural objects and natural history specimens for research or exhibit purposes to institutions in nearly every state and more than seventy countries.

The Museum's attic was very crowded until the Museum Support Center opened in 1983.

SPOTLIGHT COLLECTING FOR THE FUTURE

Meteorites collection manager Linda Welzenbach shows off one of the 924 meterorites collected in a 2003 expedition to Antarctica.

Museum botanist Robert Sims collects marine algae on a rock outcrop off the Dominican Republic.

Rick Potts, director of the Museum's Human Origins Program, records measurements of fossil bones taken at an archaeological site in Kenya.

LEARNING FROM THE COLLECTIONS

The Museum's invaluable collection of natural history specimens and cultural artifacts is a library of life. Biologists study and compare specimens to understand how species are related and how they evolved. Anthropologists examine human artifacts to document cultural continuity and change.

The anthropology collections are also an important part of the heritage of Native Americans and other cultural groups, who exchange information with Museum staff when they visit the collections. The artifacts in the collections can be a source of information and inspiration for communities that are reviving their cultural practices and artistic traditions.

Over the years, the collections have had some very specialized and sometimes surprising applications. Since the 1920s, Museum anthropologists have helped Federal Bureau of Investigation researchers identify human remains. The Museum's human skeletal collection enables physical anthropologists to assist in the identification of victims of crimes, natural disasters, and wars.

Beginning in the 1960s, military and commercial airlines have depended on Museum researchers and collections to identify the birds that collide with planes. Museum scientists take bits of beaks, bones, and feathers recovered from planes and match them to bird specimens in the collection. Once the birds are identified, airport personnel and airplane designers can take action to prevent future collisions.

When excavating burial sites like this one in Virginia, physical anthropologist Doug Owsley takes thirteen measurements of the skeleton to help determine the individual's body build and structure.

Both military and health organizations make use of the Museum's collection of 2 million mosquito specimens to identify disease-bearing mosquitoes. The collection—the largest in the world—was instrumental in the identification of the thirty different mosquito species that carry the West Nile virus.

And that's not all:

- Biologists from the U.S. Fish and Wildlife Service use the Museum's collections to study animals and plants of economic importance.

- Geologists with the U.S. Geological Survey depend on our extensive rock collection for guidance on oil and mineral exploration.

- When looking for plant species that resist pests or might improve crop production, scientists from the U.S. Department of Agriculture visit the National Herbarium, a collection of 5 million botanical specimens located at the Museum.

In these and many other ways, the Museum's collections play an important role in understanding and caring for our world. Preserving the past will benefit future generations in ways we may not imagine today.

The U.S. National Herbarium, located at the Museum, is one of the world's largest and finest plant reference collections.

SPOTLIGHT WHY DO WE NEED SO MANY BEETLES?

"If you looked at thirty tourists, they wouldn't look alike even though they all belong to the same species," explains Terry Erwin, a museum entomologist for more than thirty years. That same variation exists within an insect species. It's just harder for most people to see.

"Statistically, we need a minimum of thirty examples of a species from a given locality to study and understand the range of variation," explains Erwin. "Our collection also needs samples from many different places, stages of life, and times of year to make sure we have an accurate portrait of the species. This large sampling helps scientists study how a species is adapted to its habitat. Slightly whiter wings or longer legs may mean survival for an insect species."

Erwin, who has made more than 100 trips to collect insects in the South American rainforest, calculates that only three percent of the world's insect species have been described in the scientific literature. Collecting just from one tree, Erwin counted more than 1,700 species.

Entomologist Terry Erwin has collected about 10 million beetles during his thirty-three-year career.

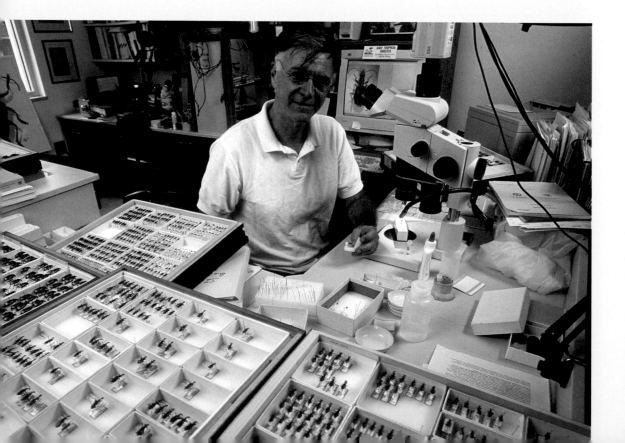

THE MUSEUM TODAY

The National Museum of Natural History is the most visited natural history museum in the world, and today's visitors can select from a rich array of learning opportunities: permanent exhibitions filled with extraordinary natural history specimens and cultural artifacts; a fascinating roster of changing exhibitions; a menu of daily presentations in the Johnson IMAX® Theater; and a variety of other educational and enjoyable experiences.

Visitors with special needs find many accommodations. All videos and media presentations within the halls are captioned, and every exhibit is wheelchair accessible. For the visually impaired, there is a special audio tour of the Rotunda, and many exhibits include touchable features.

As you tour the Museum, look for gallery hosts pointing out interesting displays and answering questions. At portable Discovery Stations, Museum educators invite visitors to touch and investigate fossils, bones, and other specimens. Daily highlights tours provide an hour-long overview of the entire Museum, while shorter tours within specific halls offer a more in-depth experience.

Staff at the Museum's two information desks will be happy to describe what activities are available on the day of your visit.

In the exhibit halls, a docent invites visitors to touch and examine real fossils.

The Museum stands only a mouse click away from all the computer terminals in the world with the most up-to-date information about exhibit openings and closings, film schedules, and special events. Teachers and home educators can also find classroom materials, curricula, electronic activities, and exhibition tours on the site—all matched with national science standards. For the major halls, associated websites offer expanded explanations, education materials, and entertaining interactive activities.

In addition, the Museum's seven science departments maintain websites that incorporate online exhibitions, collections information, research publications,

Like the black bear, every known species will have a page on the Encyclopedia of Life's website.

and recent activities. In coming years, the Museum will bring more and more of its collection databases online for the use of researchers worldwide.

The Encyclopedia of Life (EOL), a partnership of Harvard University, the Field Museum, the Marine Biological Laboratory at Woods Hole, and other institutions, is headquartered at the Museum. Eventually, it will provide a web page for each of the 1.8 million known species on earth, along with scientific literature and educational materials. To learn more, go to www.eol.org.

BEHIND THE SCENES

The world's largest group of scientists dedicated to the study of natural and cultural history works here, and their research underlies every exhibit developed at this Museum. Scientists from the Museum's major scientific departments—systemic biology, anthropology, and mineral sciences—work closely with exhibit developers, writers, designers, fabricators, and other dedicated professionals from the public programs department to plan exhibits and related programs that are both scientifically sound and educationally accessible.

Here and on the next page are samples of what may be happening behind the scenes on the day you visit.

Preparators Pete Kroehler, Steve Jabo, and Fred Grady (left to right) restore the historic *Triceratops* skeleton. Visitors were able to watch them at work in the Museum's FossiLab.

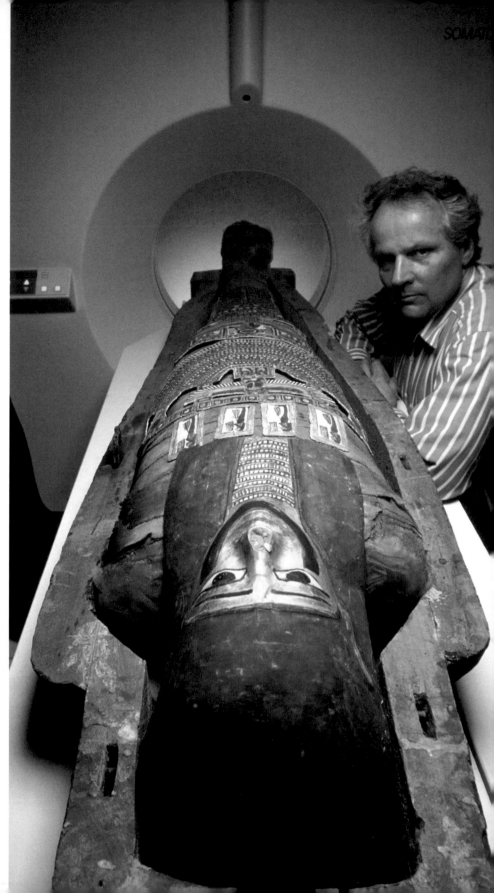

Above: For several decades, volunteers have patiently sifted through thousands of pottery fragments uncovered in Tell Jemmeh, an archaeological site in Israel, and pieced them together with great skill.

Physical anthropologist Bruno Frohlich prepares a mummy for the Museum's CAT scanner. More than 10,000 scans have helped Museum scientists safely examine fragile specimens.

WONDERS TO COME

The Museum's staff is eagerly planning a new generation of exhibits that will meet the needs of twenty-first-century visitors, keep pace with new and extraordinary advances in science, and combine the latest educational technologies with inquiry-based learning. You can sample what the future holds in some of the Museum's newer exhibits.

The amazing diversity of displays and animals in the *Sant Ocean Hall* shows why the ocean is essential to all life, including yours. In the *Janet Annenberg Hooker Hall of Geology, Gems, and Minerals,* state-of-the-art fiber optic lighting helps create dazzling and informative displays. Newly restored specimens and a wide variety of interactive displays in the *Kenneth E. Behring Family Hall of Mammals* help tell the story of how mammals adapted to an ever-changing world. Multimedia experiences in the *David H. Koch Hall of Human Origins,* opening in late 2009, give visitors front-row seats to dramatic moments in early human history.

Whatever exhibits you visit, let the time spent here spark your curiosity to know more—and even inspire the budding scientists among you. At the National Museum of Natural History, where collections grow and knowledge expands, wonders never cease.

> " A finished museum is a dead museum, and a dead museum is a useless museum.
>
> GEORGE BROWN GOODE
> SECRETARY OF THE SMITHSONIAN, 1881–96

The *Sant Ocean Hall* presents the ocean as a single global system essential to all life on Earth.

FOR OPENERS

Just inside the Museum's two entrances, dramatic displays and visitor services welcome and orient you to the wonders that lie ahead.

Whether you enter the Museum from the National Mall or the bustling of Constitution Avenue, you are immediately immersed in another world. Enticing exhibits in the Rotunda and Constitution Avenue Lobby introduce you to the Museum's collections. At prominent visitor services desks, well-informed staff can answer questions and direct you to exhibits and special programs. If you need a few minutes to orient yourself, take advantage of the comfortable seating and enjoy this magnificent space.

X-ray images of fishes from the museum's collection, like these Sea Horses, *Hippocampus* sp., are used by scientists to study skeletal structure.

At the Rotunda's elephant diorama, an interpretive label invites visitors to explore the elephant's re-created habitat—the African savanna.

Below: Visitors entering from the National Mall find themselves in the Museum's grand Rotunda.

KENNETH E. BEHRING FAMILY ROTUNDA

Crossroads, meeting place, setting for the Museum's icon: The Rotunda is quite literally the heart of this Museum. For here, in this majestic setting, most visitors first glimpse the natural and cultural wonders that await them.

Presiding over this lofty space is the Museum's icon, the largest mounted specimen of the world's largest living mammal—the Fénykövi elephant. More than 4 m (13 ft.) tall at the shoulder, it towers over a slice of its native Angola. At its feet is an intricate array of plants, animals, and objects that represent the Museum's major scientific departments and demonstrate that even giants like the elephant do not exist in isolation.

Something has caught the bull elephant's eye as he strides across the savanna, with ears flared and trunk raised. Egrets, bee-eaters, and other birds flush up at his approach. If you look closely as you walk around the diorama, you'll find evidence of other life: animal tracks pressed into the mud, dung beetles rolling away the elephant's waste, a jackal returning to its den, and the rib of a million-year-old elephant ancestor. Three interpretive rails explain how the elephant came to the Museum and how it interacts with other savanna animals. And if you want to learn more, just head upstairs.

On the second-floor balcony encircling the Rotunda, you can enjoy the elephant diorama from a different perspective and explore three discovery stations on the African elephant's evolution, anatomy, and role in contemporary culture.

A cattle egret flushes up near the elephant's foot. Other birds, mammals, reptiles, and insects are also part of the elephant's world.

At the evolution station, you can examine the fossil bones of an extinct elephant. The anatomy station invites you to stand next to an elephant's column-like legs and see how you measure up. At the third station, masks, pipes, horns, and other cultural artifacts illustrate the elephant's commanding presence in African daily life and culture.

Above: Three soaring totem poles lead from the lobby to the second floor.

Our Easter Island *moai* stands sentry at this Museum entrance.

CONSTITUTION AVENUE LOBBY

A grand, two-story space lit by chandeliers, the Constitution Avenue Lobby is a treasure chest of natural and cultural wonders. Watching over the lobby is the Museum's Easter Island ancestor figure, called *moai*, first put on exhibit in 1888.

Each case focuses on a single collection—from butterflies to minerals, fossil plants to early hominid tools. Some of the specimens, like the fossil arthropods unearthed in Canada's Burgess Shale, are more than 530 million years old. Others, like the pottery by Native American Maria Martinez, are contemporary, but reflect centuries of artistic tradition.

Dazzling specimens of pyrite, quartz, opal, and topaz illustrate some of the extraordinary shapes and colors found in the *Janet Annenberg Hooker Hall of Geology, Gems, and Minerals.* Jewel-like insects—from iridescent morpho butterflies to lace-winged

skimmers—hint at what's to be seen in the *O. Orkin Insect Zoo.* The array of animal skins and skeletons in another case may seem like reminders of death, but—as you'll learn in the *Bones Hall*—they are clues to the ongoing evolution of life.

Three massive totem poles carved by Northwest Coast Indians rise majestically into the stairwell at one corner of the lobby. Nearby Smithsonian Institution Libraries present book-related topics featuring rare works from its collections or new acquisitions. In another corner, a changing display focuses on the work of Museum researchers and programs from historic Native American shirts to broken pots and endangered plants.

SPOTLIGHT OLMEC COLOSSAL HEAD

Colossal heads from Mexico's ancient Olmec culture (1500–400 BC) are some of the oldest sculptures in North America. There are seventeen known heads, many of which appear to be portraits of Olmec leaders in ball-playing headgear. The heads were carved without iron tools and moved before the wheel came to North America.

A replica of colossal head No. 4, carved by Mexican sculptor Ignacio Pérez Solano, sits near the Museum's north entrance. The original was excavated in 1946 by Smithsonian archaeologist Matthew W. Stirling at San Lorenzo Tenochtitlán and is on view in Veracruz, Mexico.

A replica of Olmec colossal head No. 4 greets visitors entering from Constitution Avenue.

Africa

Welcome
Mammal Fam
Come meet yo

BIOLOGY HALLS

From tiny ants to towering giraffes, the Museum's biology halls present the beauty and wonder of animals living today and explore the forces that led to this diversity.

Compared to the Earth presented at the beginning of the Fossils Halls, our planet today is rich in many kinds of life-forms. The mission of biologists at the National Museum of Natural History is to discover, name, and classify the natural world, and to explain the evolutionary and ecological processes that created this diversity. The biology exhibition halls, located on all three floors of the Museum, reflect this mission.

A pouncing tiger is one of the 274 species of mammals that greet visitors to the *Kenneth E. Behring Family Hall of Mammals.*

SANT OCEAN HALL

A spectacular array of echinoderms, such as sea stars, sea urchins, and sea lilies, from the Museum's vast collections are found in the Ocean Hall biodiversity display.

Opposite: The *Sant Ocean Hall* introduces visitors to the ocean as a dynamic global system that is essential to all life—past, present and future.

You live on an ocean planet. Looking down from space, the ocean appears as a magnificent swath of blue circling the globe. It spans many vast basins, forming one global system that's essential to all life on our planet—including yours.

The ocean shapes Earth's climate and drives the weather, yielding as much oxygen as all of the world's forests combined. It provides food and livelihoods for many peoples from around the world. Still, more than 95 percent of the marine world has yet to be explored.

In the *Sant Ocean Hall*, you'll discover the ocean in all its complexity and unearthly beauty. Our unparalleled collections and ongoing research anchor the exhibition, the largest ever developed by the Museum. Each gallery weaves together an amazing tale featuring an incredible diversity of marine organisms and unique ecosystems that lie beneath the surface of the sea.

Teeming with life, the *Sant Ocean Hall* will inspire you to make connections between the ocean and your own life, to continue your exploration of the ocean, and to become a better steward of this vital part of our life support system.

DIVE IN

To greet visitors, the Ocean Hall presents its official ambassador, a spectacular model of a North Atlantic right whale that gracefully descends through the two-story atrium. Three magnificent tree-of-life displays introduce Earth's immense diversity with representatives from every known phylum, from the tiniest to the titanic. Don't miss a nearly mythical creature of the deep—a 7.5 m (24.5 ft.) female giant squid caught in a fisherman's net off the coast of Spain.

Exploring humans' age-old connection to the ocean, a treasure case presents artifacts from around the world, including a classic surf board, an Easter Island dance paddle, a Yup'ik horned puffin mask, and a model of a modern container ship. Look for more stories that highlight our impact and dependence on the ocean throughout the hall.

JOURNEY THROUGH TIME

Opposite: Along a warm, inland sea during the Cretaceous Period, rudist clams dominate the reefs.

How did the ocean become what it is today? The story begins 3.5 billion years ago when the first life-forms—tiny, single-celled microbes—appeared in the sea. Another two billion years would pass before enough oxygen built up in the air and water to support more complex life.

Then, in a sudden surge of diversity about 600 million years ago, many new groups of animals evolved. The Burgess Shale fossils, first discovered by a Smithsonian scientist, tell the important story of the Cambrian Explosion. Formed by an underwater mudslide in what is now Canada, rare fossils of both soft- and hard-bodied marine organisms offer a glimpse of this ancient world.

As ocean ecosystems changed over time, species rose and fell in bursts of adaptation and extinction. Murals drawn from the fossil record dramatically envision what once lived in the ocean. Some 350 million years ago forests of feathery crinoids ruled the shoals. By 100 million years ago rudist clams dominated the scene. The era crashed to a close 65 million yeas ago in the aftermath of a colossal asteroid impact that wiped out almost three-quarters of all marine species—along with the dinosaurs.

Long spines kept this trilobite from burrowing in the sand to hide, but they also protected it from predators.

Above the fossil whale display, three skeletons chart the whale's evolution from ancestral land mammal to streamlined denizen of the sea, ushering visitors into the vast open ocean ahead.

MEET PHOENIX!

A 988 kg (2,200 lb.) model of an actual North Atlantic right whale reigns over the hall's grand space. Accurate in every detail, she's big and beautiful! She's one of about 400 of these whales left in the world. Tracked since birth, Phoenix and her story link every corner of the Ocean Hall. Her big size means a big appetite: some 2,200 pounds of krill every day! In the displays below the model, you learn more about how Phoenix and her kind live. An array of hunting and ceremonial artifacts from Arctic indigenous communities reflects their respect for the whale and its gifts of food, fuel, and bone.

Centuries of commercial whaling and other environmental factors have brought whales to the brink of extinction. Now heroic conservation efforts protect this highly endangered species.

The North Atlantic right whale model hanging from the Ocean Hall's ceiling is a precise replica of "Phoenix," a 13.7 m (45 ft.) long living female whale. She has been tracked by scientists since her birth in 1987 as she migrates seasonally between calving grounds and feeding grounds along the eastern coast of the United States.

OPEN OCEAN AND OCEAN EXPLORER THEATER

Take a seat in the theater, and plunge through the water column in a manned submersible to the most remote habitat on Earth, the bottom of the sea.

In the "Open Ocean" displays nearby, you'll explore the three distinct zones. Only the thin surface waters receive enough light for photosynthesis—the heart of the ocean's food web. Just 201.2 m (660 ft.) down in the twilight zone, sunlight and food are scarce, and many species engage in the largest vertical migration on Earth as they commute at night to the top to feed. An amazing variety of species have adapted to the dark by producing their own light, known as bioluminescence. Despite icy cold, eternal darkness, and crushing pressure, the deepest zone—the ocean floor—holds a dizzying array of species, many with very unusual adaptations.

"Science on a Sphere," created by the National Oceanic and Atmospheric Administration, uses a dramatic multimedia presentation to explain many of the complex aspects of the ocean

GLOBAL OCEAN SYSTEMS

The "Science on a Sphere" display offers an unusual venue, the 1.8 m (6 ft.) diameter globe that highlights data from satellite observations to demonstrate how ocean systems function as one large global system. Visitors see a dynamic ocean producing our food and oxygen, driving Earth's weather and climate, and dramatically changing the face of our planet throughout its history in the media experience, which was developed by the National Oceanic and Atmospheric Administration (NOAA).

In the surrounding displays, visitors hear from oceanographers and find out about the specialized tools needed to delve into the deep. The Ocean Today kiosk features current ocean news, stories about the ocean and its creatures, and new research.

REEF, SHORES AND SHALLOWS, POLES, AND COLLECTIONS

Standing on the shore, looking out over a sparkling expanse of sky and waves, the ocean seems familiar, inviting. Coastal waters are our gateway to the ocean, the place where humans have the greatest impact on its ecosystems. In fact, nearly 50

Oceanic communities at 21.4 m (70 ft.) at Carrie Bow Cay Research Station, Belize.

percent of the world's population lives within 90 miles of the shore.

Coasts can be rocky or sandy, icy or muddy. Visitors have the opportunity to compare different types of coastal ecosystems within a few steps. A cross-section of a Chesapeake Bay estuary reveals that ecosystem's amazing diversity, and another diorama pictures the amazing number of animals that live just below your beach blanket, wedged between grains of sand. Nearby, you'll find another exceptional marine ecosystem—a live Indo-Pacific reef where visitors can witness dozens of ocean organisms in a 5,700 l (1,500 gal.) aquarium.

At the ends of the Earth, animals adapt to similar conditions—year-round cold, months of darkness, shortages of food—yet there are profound differences between the two communities at the poles. The Arctic Ocean, nearly landlocked, is relatively calm. Gray whales and walruses ply its waters, and polar bears ride ice floes to hunt for seals. The Southern Ocean, with no surrounding landmass, is far more turbulent and supports more abundant and varied species.

The marine collections display holds a vast library of life—specimens, X-rays, videos, and behind-the-scenes views of research. Museum scientists and their colleagues from around the world rely on a collection of nearly 30 million specimens—and growing—as they work to unravel the mysteries of evolution.

This canoe was carved by a Tlingit craftsman especially for the *Sant Ocean Hall.*

LIVING ON AN OCEAN PLANET

How are humans impacting the sea? How is our own future tied to the health of the ocean? In this gallery, a more distinctly human perspective calls on visitors to join in a new era of ocean stewardship. Computer stations challenge users to think about the choices to be made if we are to manage marine resources and confront climate change. In addition, an updatable video and photo display highlight current ocean research around the world.

SALMON AND PEOPLE

Discover how a fish changed the way people lived in the North Pacific. About 18,000 years ago, salmon began migrating up freshwater streams and rivers to spawn. As salmon flourished, so did people. With an abundance of salmon, early hunting and foraging societies began to settle into villages which expanded their social networks. Once settled, they hosted elaborate festivities—the inspiration for many artistic traditions. An example of one tradition hangs overhead: an 8 m (26 ft.) carved canoe commissioned by the Smithsonian and carved in Alaska by a member of the Tlingit tribe.

Coral reefs—found in warm, clear, shallow waters—support a rich diversity of marine life, such as these anthiine seabasses from the Red Sea.

PARTNERS IN EVOLUTION:
BUTTERFLIES + PLANTS

It's a picture-perfect summer day inside the Museum. The light is bright, the air warm, and butterflies perch delicately on beautiful blossoms.

Hundreds of butterflies flutter from flower to flower in the cocoonlike butterfly pavilion. Frequent bursts of steamy tropical air and 27° C (80° F) temperatures make it feel like the tropics all the time. Outside the chamber, dramatic images and interactive displays weave together the complex story of butterflies and plants, partners in evolution for tens of millions of years.

A rose swallowtail butterfly in profile.

WELCOME

A moth with a 20.5 cm (8 in.) long tongue? An orchid with a throat to match? It's no accident that the two have such a close fit. For thousands of generations, the giant hawk moth and the Madagascar star orchid have been changing in response to one another. In *Partners in Evolution: Butterflies + Plants*, hundreds of butterflies, moths, and plants bring their age-old connection to life.

LIVE BUTTERFLIES
butterflies.si.edu

The exhibit houses the first year-round, indoor butterfly habitat in the Washington area. You can reach the exhibit from the *O. Orkin Insect Zoo* or the second floor Rotunda.

Before entering the habitat, take a moment to look at the display along the wall and see how butterflies and plants interact—sometimes as friends, and sometimes as foes. Hungry caterpillars can devour plants, but some plants keep them at bay with sharp hairs and sticky substances that stop caterpillars in their tracks. However, plants enlist butterflies' help, luring them with sweet flower nectar. Butterflies unknowingly carry pollen from flower to flower, enabling plants to reproduce. Butterflies and plants could not exist without each other.

INSIDE THE BUTTERFLY PAVILION

Small chrysalides arrive in weekly installments year-round from special farms in Africa, Malaysia, and Central and South America. In the Transformation Station, their nugget-like bodies hang as they would in nature while the caterpillars' tissues reorganize and adult butterflies take shape.

For two to four weeks—an average adult butterfly lifetime—the butterflies flit from one blossom to another, sipping nectar, roosting, and flexing their wings to warm up before they fly. Several dozen different species flutter across the milky dome, from iridescent blue morphos to zebra-striped kites and brilliant scarlet swallowtails.

At the feeding stations, butterflies alight on chunks of grapefruit, banana, and other fruits. Watch carefully to see them uncoil their straw-like tongues and sip nectar or other sweet liquids from trays. Some butterflies appear to be just standing on the fruit; in fact they're tasting it. Their taste buds are on their legs!

A monarch butterfly perches on a pink zinnia.

Moths and butterflies come in a stunning variety of colors and patterns, which helps protect them from predators. Some, such as the Sphinx moth, match the bark where they roost, while others like the Monarch telegraph their bitter, or even poisonous, taste with conspicuous wing patterns. At a glance, eye spots on the wings of the owl butterfly may look like the eyes of a predator and frighten away birds and lizards.

EVOLUTION REPEATS ITSELF

How is a butterfly like a hummingbird? Or a bee? Or a bat? All of them have long mouthparts for gathering plant nectar, and all pollinate the plants on which they feed. Though unrelated, these animals co-evolved with plants in surprisingly similar ways—a process called convergent evolution.

In the display, five very different animals with similar diets and ways of gathering food show the process at work. Palamedes swallowtail butterflies siphon nectar with strawlike tongues, and purple-throated carib hummingbirds insert long, curved beaks into slender flowers to gather nectar. Tiny tanglevein flies lap up nectar with spectacularly long tongues, and endangered lesser long-tongued bats drink from the giant saguaro cactus' narrow blossoms. With their long tongues, orchid bees feed from deep-throated flowers in the American tropics.

This Ulysses swallowtail butterfly is native to Australia.

FRIENDS OR FOES?

Plant-butterfly partnerships have been evolving for millions of years. Several intriguing case studies flanking the windows illustrate how some plants arm themselves to the hilt while others evolved mutually beneficial connections.

Take the yucca moth and the yucca plant. For some 40 million years, yucca moths have played an active role in fertilizing yucca plants. The moths deposit their eggs in the yucca flower's ovaries, then pack pollen beneath their heads to deliver to the stigma of the next plant they visit.

Milkweed plants defend themselves from insects with milky sap that sticks to mouthparts and bodies. Not to be foiled, milkweed beetles have evolved a behavior in which they chew cuts into leaf veins, letting the latex drain out so they can feed.

As caterpillars turn into adult butterflies, they change their relationship with plants, becoming friends instead of foes. Finally, the leaf-eating stops when a caterpillar sheds its skin like exoskeleton for the last time, transforms to a chrysalis, and then to a butterfly in a magical process called metamorphosis. A time-lapse video captures this dramatic process at work.

CHANGING WORLD, CHANGING SPECIES

It's taken tens of millions of years of evolution to produce the amazing diversity of butterflies and moths seen today. Along the way, many species disappeared, while others endured. Some gave rise to multiple new species, including the specialized day-flying moths we now call butterflies.

Banded orange butterflies are found from Brazil to central Mexico.

In this display, a series of murals and rare plant and insect fossils paint an image of the world during four distinct time periods. The story begins 170 million years ago when there were no flowers and only a few major groups of moths with chewing mouthparts. The first blossoms finally appeared by 102 million years ago, and by 48 million years ago butterflies had appeared on the scene, imbibing nectar and other fluids.

Today there is an explosion of diversity that includes more than 350,000 species of flowering plants and over 150,000 kinds of moths and butterflies. As if to echo the flood of new species, a swirl of butterfly images lifts off from the exhibit rail and rises to the ceiling with a gentle flutter of wings.

SCIENTISTS AT WORK

How do we know so much about the partnership between butterflies and plants? Museum scientists examine evidence from the field and the Museum's vast collections. A display near the Rotunda entrance gives you a sense of just how large the Museum's plant, insect, and fossil collections are, and the critical role they play as scientists continually unravel the mysteries of evolution.

KENNETH E. BEHRING FAMILY
HALL OF MAMMALS

About 210 million years ago, when all the land on Earth was joined in one giant supercontinent, an unprepossessing animal appeared beneath the footsteps of early dinosaurs. It was small enough to fit in your hand and rather unimpressive. But it represented a milestone in the history of life on Earth: It was the first mammal. From this common ancestor, tens of thousands of mammal species evolved— including more than 5,000 kinds living today. From this tiny animal, you evolved.

Reopened in the fall of 2003 after a major renovation, the Mammals Hall presents the wondrous diversity of mammals and shows how they adapted to a changing world. Designed with families in mind, the interactive exhibit includes four discovery zones rich in hands-on activities as well as a fast-paced theater presentation on mammal evolution.

If you really want to get to know your relatives, the Museum's Mammal Hall is the place to go.

Opposite: High in a tree, a leopard guards its next meal. Beyond, an aoudad surveys the desert habitat in the "Africa Gallery."

A giant anteater, with baby on board, laps up termites on the rainforest floor— part of the "South America Gallery."

In a dramatic display just outside the Evolution Theater, the spotlight is on tiny *Morganucodon oehleri*. Less than 10 cm (4 in.) long, *Morganacudon* lived about 210 million years ago. It wasn't the first mammal—no one will ever know exactly who that was—but it was a very close relative. *Morganacudon* lived in the shadow of the early dinosaurs, and probably came out to hunt for food at night, when it was easier to escape potential predators.

Morganacudon's close relative, the first mammal, passed on its DNA to millions of descendants—including you. You can step inside a display where special lighting effects show our common ancestor's genes being passed on to an evolving diversity of mammals. Or visit the Evolution Theater, where *Morganacudon* is the star of the rousing show.

This model of *Morganucadon oehleri* shows what the first mammal may have looked like.

ORIENTATION GALLERY

Welcome to the mammals family reunion! As you enter the hall, you are surrounded by more than sixty of your relatives. A tiger leaps for its prey, a sloth hangs by its limbs, and a giant panda munches on bamboo. Bats fly by, and a manatee mother and her young swim overhead. A white rhinoceros collected by President Theodore Roosevelt shows how large mammals can get. A tiny European mole represents the other extreme.

What do all these animals have in common? Hair, for one thing. It may be soft and thick like the sheep's wool, sharp like the porcupine's quills, or as sparse as the hair on the walrus. But at some point in their lives, all mammals have hair on part of their bodies. Milk is another characteristic unique to mammals. All mammal mothers produce milk to feed their young.

A third feature that unites all mammals past and present is less visible: special ear bones. Over millions of years of evolution, two bones in the jaws of our reptile-like ancestors moved to the middle ear. They became the "hammer" and the "anvil"—two parts of a sound-amplifying system found only in mammals.

How did so many different shapes, sizes, and behaviors evolve from a single ancestor? The world has changed drastically over the past 210 million years. As the world's climates and habitats changed, mammals had to either adapt...or go extinct. To see the different ways mammals adapted to this changing world, you'll need to travel to four continents.

The Argali sheep (top) and the Chinese water deer are just two of the diverse mammals that greet visitors entering the hall.

AFRICA

A male lion stands sentry at the entrance to this gallery, and two females take down a buffalo nearby. It's the end of a long, hard dry season. Suddenly, thunder rumbles, the sky darkens, clouds roll in, and lightning streaks across the sky. A unique media presentation immerses you in the alternating seasons of rain and drought to which African mammals have adapted.

Today, Africa is largely savanna and desert. But 15 million years ago, lush rainforests stretched across the continent. Giraffes had short necks. You can touch a cast of a neck vertebra from a 14-million-year-old extinct giraffe, and compare it with that of the modern giraffe. The tallest living mammal, today's giraffe stands up to 5.8 m (19 ft.) tall on stilt-like legs and is adapted for browsing in open areas.

SAVANNA

As forest gave way to savanna, African mammals adapted in different ways. Wildebeest have specialized teeth and a complex digestive system for eating tough

One female lion claws a buffalo's back while another avoids its thrashing hooves. By cooperating, lions can bring down an adult buffalo.

grasses. An animated video follows the journey of this fibrous food through the wildebeest's body—and shows what comes out at the end.

Waterholes are critical for survival on the savanna. During the dry season, a single waterhole may be the only place to drink for miles. In the waterhole display, a giraffe spreads its legs wide to lower its long neck and drink—a posture that would cause blood to rush to the giraffe's brain if it hadn't evolved special neck valves to regulate blood flow.

Nearby, on the floor, are the 1.5-million-year-old footprints of another mammal. They belong to one of your early human relatives—either *Homo erectus* or *Paranthropus robustus.* A scientist at this Museum and one of her Kenyan colleagues found the footprints in 1978.

SAHARA DESERT

Food and water are scarce year-round in the world's largest desert, yet more than ninety kinds of mammals have evolved ways to live here. How do they survive in this bone-dry land, where daytime temperatures can reach 55°C (131°F), and plants are few and far between?

In the shadowy forest, the markings on the highly endangered bongo help it vanish before your eyes.

The scimitar-horned oryx has long legs that raise its body above the hot sand and a white belly that reflects heat rising from the ground. And while a human needs 6 l (1.6 gal.) of water a day to survive in the Sahara, an oryx drinks none! It feeds at night, when plants have absorbed the evening dew, and its highly efficient kidneys retain almost all the water it takes in.

Fennecs spend the day in their burrows, where the temperature is forty-five degrees cooler than at the surface, and venture out to hunt only at night. The enormous ears of these tiny foxes work like funnels to collect the sounds of prey.

RAINFOREST

Finding water is easy in the lush African rainforest. But moving around and staying in touch are more challenging. Many large mammals, like the okapi and bongo, have dark, striped coats that help them blend into the forest shadows. Their bodies are smaller and more compact, and their horns smaller and closer to the body, than those of their relatives living on the open savanna—helping them squeeze through tight places.

Ever heard a bushbaby? A sound interactive enables you to listen to the amazing sounds that these tiny primates make in the dense forest. Colobus monkeys, on the other hand, depend on body postures and striking black-and-white markings to communicate with each other.

From the frigid, windswept tundra to sun-dappled forests, North American mammals weather seasons of warmth and cold, cycles of plenty and scarcity. Explore how these mammals have adapted to seasonal changes in three different habitats.

With the warmest coat of all terrestrial mammals, this Arctic fox is toasty even at temperatures of –40°C (–40°F). Its furry tail insulates its nose during naps.

FROZEN NORTH

Mammals of the Far North face long, freezing winters and snow-covered landscapes. Many have evolved body shapes that conserve heat, feet that can traverse or dig through snow, and white coats that provide winter camouflage. An interactive demonstrates how mammals with white fur disappear in the snow, making it easier for them to hide.

The caribou has dense underfur and stiff guard hairs that provide insulation, and shovel-shaped hooves that can scrape away snow to find food. When winter approaches, many caribou herds migrate up to 1,000 km (621 mi.) to reach their forested winter home.

The sound of howling winds and creaking ice at the Frozen North Discovery Zone may make you shiver. Inside, you can find out how polar bears stay warm all winter, how seals survive in the frigid seas, and what keeps the sea otter warm and dry. Touch the refrigerated model of a ground squirrel to feel how low its temperature drops during hibernation—just above freezing!

WIDE OPEN PRAIRIE

Across vast seas of grass on North America's prairie, some mammals run like the wind while others dart underground. Small mammals like the prairie dog escape extreme weather and predators by living underground. They evolved sharp claws for digging and long body shapes that can squeeze through narrow tunnels. A cutaway prairie dog burrow includes a nesting chamber, latrine area, turnarounds, and plugs to deter ferrets.

Unlike prairie dogs, large mammals like the bison and pronghorn cannot hide on the open prairie. For them, running is the best defense. The fastest mammal in North America, the pronghorn can reach speeds of 105 kph (65 mph). The larger, sturdier bison is not as fast, but it can still reach speeds of 89 kph (55 mph). A short video compares these two prairie runners.

TEMPERATE FOREST

One of North America's oldest habitats, the temperate forest once stretched across the continent. Trees and shrubs in these seasonal forests shed their leaves every fall and sprout anew in the spring. Forest mammals give birth when spring's new growth heralds a season of plenty, and scurry to store food in the fall.

In the spring forest display, you'll see mammal babies galore—including a black bear cub, raccoon cub, fox kit, and white-tailed deer fawn. A video shows how mammal mothers teach their young to find food, protect themselves from danger, and negotiate the temperate forest.

Above: Black-tailed prairie dogs live together by the thousands in extensive tunnels. This one is scanning the prairie for potential enemies.

The Canada lynx's big feet keep it from sinking into snow when it ambushes its favorite prey— the snowshoe hare.

In the Temperate Forest Discovery Zone, the mammals have been busy collecting and storing food for the lean months of winter. You'll discover some surprising items in the forest pantry.

A beaver sits snugly inside its lodge, safe from the coyote sniffing above. Walk around the lodge to discover the beaver's winter food supply— cottonwood, willow, and oak logs. Beneath a mural of a colorful autumn forest, you can hunt high and low for hidden food caches— earthworms for the Eastern mole, snails for the short-tailed shrew, nuts for the chipmunk.

North America's largest living carnivore, the brown bear looks similar to its 250,000-year-old ancestors.

Nearly 65 million years ago, rainforests covered most of South America. Today, a descendant of one of these ancient forests—the Amazonian rainforest—survives as the world's largest tropical rainforest. Nearly 200 kinds of mammals live here, more than in any other habitat on Earth. They live high and low, feast on different foods, and take day and night shifts.

SHADY RAINFOREST FLOOR

The almost continuous canopy of evergreen trees shades the rainforest floor, stifling the growth of young plants. So what do these mammals eat?

Two rodents—the agouti (top) and the capybara (bottom)—crack nuts and clip grass, respectively. The capybara is the world's largest living rodent.

Insects, for one thing. The giant anteater laps up thousands of ants and termites in a single day. Its sickle-shaped claws tear through tough nests and logs, its long muzzle probes deep into insect tunnels, and its saliva-coated tongue moves in and out 160 times a minute to lap up fast-fleeing prey. A touchable model of a giant anteater skull reveals that it is toothless, and an animated video explains how this mammal makes fast work of insects despite its lack of teeth.

SUN-FILLED RAINFOREST CANOPY

In the crowded canopy, mammals avoid fighting over food by eating different things. The three-toed sloth hanging from a branch with its hooklike claws is a leaf lover. Leaves aren't very nutritious, but slow-moving sloths are experts at conserving energy. Two extra neck vertebrae enable them to swivel their heads nearly 270 degrees without moving their bodies.

The world's smallest monkey, the pygmy marmoset is a sap sucker. It feeds on sugar-rich tree saps and gums, as well as protein-filled insects. The marmoset's tiny skull is on display beneath a magnifier so you can see how its teeth are adapted for chiseling tree bark.

WHO WORKS THE NIGHT SHIFT?

Sounds of the night surround you as you enter the Rainforest Discovery Zone. After sundown, a different crew of mammals runs the Amazonian rainforest. Discover how these mammals use darkness to their advantage.

Left: Because of its owl-like hoots and large eyes, this Northern night monkey is sometimes called an owl monkey.

A jaguar is about to pounce on an unsuspecting paca. Furred foot pads muffle the sound of the cat's footsteps, and long whiskers feel objects in the dark. It sees six times better at night than you do. The paca is a goner.

Nearby, you can look into viewing portholes to see the eyeshine of three nocturnal mammals. Depending on eye pigment, eyeshine comes in a variety of colors. Use the flashlights to look for the paca's yellow eyeshine, porcupine's red eyeshine, and kinkajou's green eyeshine.

Unlike the jaguar, bats see in the dark about as poorly as you do. To find their way around at night, they evolved a specialized sound system called echolocation. A computer interactive explains how it works and challenges you to catch as many moths as you can in one minute. Nearby is a bat that is out for blood—the vampire bat, which makes a living feeding on the fresh blood of other unsuspecting animals.

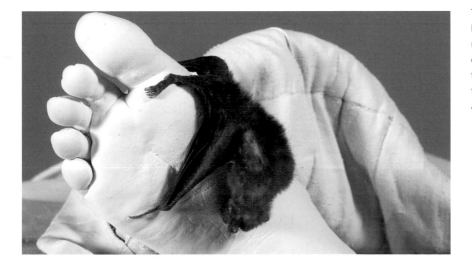

To survive, vampire bats need to lap up nearly 25 mL (5 tsp.) of fresh blood every two days. This one is feasting on the foot of a sleeping human.

AUSTRALIA

Kangaroos and koalas, the duck-billed platypus and the echidna—Australia is home to a fascinating variety of mammals found nowhere else in the world. Discover how they hop, glide, and fly across this island continent.

LAND OF KANGAROOS

The two kangaroo species on display look very different from each other. That's because long ago, their ancestors adapted to very different habitats. The red kangaroo is a creature of the open grassland, built to hop. Long hind feet push it off the ground, and powerful back limbs propel the animal forward. A touchable model of a kangaroo's foot shows how it is designed to take a pounding, and a short video shows a kangaroo hopping normally and in slow motion.

The other kangaroo species on exhibit is up a tree—literally. The original tree kangaroos moved into the rainforest canopy for new food sources, and their descendents evolved legs and feet suited to climbing. A cast of a 24-million-year-old fossil kangaroo shows that its legs and feet resembled those of modern tree kangaroos—evidence that it was probably a climber.

The red kangaroo hops for miles at 25 kph (15.5 mph). In an emergency, it bounds away at twice that speed.

GETTING AROUND IN OPEN WOODLANDS

Patches of dry woodlands dominated by eucalyptus trees ring Australia's interior, and the mammals there have found ways to meet the challenges of life on the limb. Some glide from tree to tree. Sugar gliders swoop across the case, illustrating the various stages in gliding. These specialized possums have fur-covered membranes between their fore and hind limbs to catch the air.

Bats are the only mammals that fly, and several different species take wing in this case. Koalas climb. They spend nearly their entire lives in eucalyptus trees, dining on leaves that are toxic to most other species. A life-sized model of a koala's hand shows the needle-sharp claws, opposable fingers, and ridged pads that help these marsupials get a grip in the trees.

A red-necked wallaby spends its first nine months in the pouch. Even after it's too big to fit, it returns to nurse for several more months.

ONLY IN AUSTRALIA

In the Australia Discovery Zone, you'll meet the three major groups of living mammals: monotremes, mammals that hatch from leathery eggs; marsupials, mammals that grow up in pouches; and placentals, mammals that develop in a womb, nourished by a placenta. Because of Australia's long history of isolation, it is the only continent where all three groups have survived to the present day.

There are only two kinds of monotremes in the world today, the platypus and the echidnas, and you can see both here. You can even peek into a platypus burrow and look for a mother incubating her eggs. Marsupials are represented by a red-necked wallaby with a joey in her pouch, and placentals by a red fox mother and young. Three amazing videos show a newborn from each of these mammal groups coming into the world and nursing.

From the common ancestor of all mammals, different groups of mammals evolved. A large case presents three major families of placental mammals. Each display juxtaposes specimens of modern mammals with photographs of their extinct ancestors and touchable fossil casts.

Wolves, seals, and other carnivores are the ultimate mammal predators. A cast of a 49-million-year-old fossil carnivore, *Paroodectes feisti,* shows that it already had the shearing teeth that define carnivores.

Ungulates evolved their grinding teeth and hoofed feet to eat tough plants and flee from fierce predators. The first ungulate, *Phenacodus primaevus,* was up and running 60 million years ago, and it already had long limbs and stubby hooves.

Watch out! The striped skunk's handstand warns of an impending spray of noxious mist—this small carnivore's way of defending itself.

Every primate—including you—shows evidence of its tree-dwelling ancestry: grasping hands, 3-D vision, and complex brains. *Smilodectes gracilis,* our 50-million-year-old relative, had these features, too.

Inside the Evolution Theater, a lively presentation dramatizes a reunion of mammals past and present. Take a seat and meet some more fascinating relatives.

With long, clawed fingers, golden lion tamarins pick insects and debris from the coats of relatives.

Bird-watcher or not, there's no need for binoculars here. This small exhibit outside Baird Auditorium provides a close-up look at nearly 500 species of birds recorded in the area around Washington, D.C.

A bald eagle and a pair of rare transients—golden eagles—stare down from a rocky outcrop. Facing them are thirteen species of hawks, including the endangered peregrine falcon—occasionally seen nesting on Washington office buildings. Stroll past two wild turkeys pecking for food, eight species of owls, and a case filled with dozens of tiny warblers perched on branches. Or make a stop at the woodpecker case to admire two male pileated woodpeckers—at 42 cm (16.5 in.) long, the largest in this area.

Waterfowl and shorebirds are well represented by cormorants, loons, ducks, grebes, herons, egrets, and more. The different bills and feet illustrate how they evolved for finding food in water. Tucked among the terns is a rare fall migrant— the red phalarope. A pair of preening tundra swans ends the exhibit.

A Northern goshawk feeds on a blue jay.

As you explore, keep an eye out for three birds that you will not find in the wild no matter how powerful your binoculars, for these species are now extinct: the passenger pigeon, Carolina parakeet, and heath hen. The two passenger pigeon specimens on display were obtained in the old Washington Market in 1860, when billions of these colorful birds still filled the skies. The last passenger pigeon died in Ohio in 1914.

O. ORKIN INSECT ZOO

Insects may be tiny. But together, the world's insects outnumber humans by 20 million to one! In the *O. Orkin Insect Zoo,* you can get eye-to-eye with live insects and other arthropods and marvel at the adaptations that made them so successful.

WELCOME

At the entrance to the Insect Zoo, dozens of preserved arthropods illustrate the diversity of this large group. You'll find the world's heaviest insect (the Goliath beetle from West Africa) and a walkingstick with a wingspan of 33 cm (13 in.). Like all arthropods, they have bodies made up of segments and an exoskeleton—a shell-like external covering.

Insects and other arthropods have been amazingly successful—they're been around for over 475 million years. A diorama provides a glimpse into a swamp from

A colorful mural depicting a variety of arthropods greets visitors to the popular *O. Orkin Insect Zoo.*

the Carboniferous Era, 300 million years ago, where ancestors of our modern cockroaches forage among rotting plant remains.

THRIVING THROUGH CHANGE

A talent for adaptation is another reason for insects' phenomenal success. A display highlights the ingenious eating mechanisms that enable insects to feed on a wide variety of foods. Mosquitoes pierce the skin of plants and animals with needlelike tubes, while butterflies use strawlike tubes to sip sweet nectar from flowers.

A display of live crickets—lots and lots of crickets—illustrates how insects' speedy passage from birth to adulthood enables them to increase their numbers very rapidly.

DEALING WITH DANGER

Insects are small and surrounded by danger, but they protect themselves with an impressive range of weapons. Some avoid danger through camouflage. See if you can pick out the Australian stick insects from the twigs on which they're perched, or the leaf insects from the surrounding foliage.

This sunset moth is one of dozens of preserved insects that provide an opportunity for close examination.

Other insects warn enemies to stay away. Milkweed bugs, for example, use bright red and black colors to alert birds and other predators: "We taste *bad.*" Leaf-footed bugs secrete a smelly substance when disturbed.

SPIDER STRATEGIES

Without spiders, insect populations would be totally out of control. Wolf spiders are active hunters, using their keen eyes to stalk insect prey. Other spiders just sit and wait for dinner. A series of illustrations highlights the incredible variety of silken tangles and sheets web-spinning spiders use to trap their prey.

INSECT SOCIETIES

Termites, ants, and many bees and wasps band together in highly organized colonies to work for the greater good. Peek inside the underground nest of the desert-dwelling honeypot ants and admire the glistening golden globes suspended from the ceilings. These globes are actually worker ants full of regurgitated nectar, serving as living storage jars from which other honeypots draw honey.

Visitors of all ages can crawl inside a life-sized model of a termite mound. It is fashioned after a nest of the African mound termite, a marvel of engineering that can rise to 5.49 m (18 ft.) in height—a fitting castle for the termite queen and king and their 2 million subjects.

At the Insect Zoo, you can get close enough to see the hairs on a tarantula. Tarantulas use these hairs to locate prey.

Within this re-created home live a host of imaginary insects. Visitors can push buttons to find out where they're hiding.

Have you ever wondered how a honeybee hive works? Through a unique cut-away window, you can watch honeybees flying to and from a real hive. A tube leads outside, where bees collect nectar from flowers in the Smithsonian's gardens during the warm months.

OUR HOUSE, THEIR HOUSE

To an insect, our homes and backyards make excellent habitats. Whether we bring them in ourselves or they come on their own, insects find their way to places all over the house. At a re-creation of a home, you can play a game to find out which insects live where.

Silverfish? Look in the bookcases, basement, and kitchen. Flour beetles? They're feasting on grains, pastas, chocolate, and more in the kitchen. And then there are the clothes moths, carpenter ants, fleas . . . and, of course, the cockroaches. If left alone, one female German cockroach can produce over 30,000 offspring in a single year!

AT HOME EVERYWHERE

Arthropods have adapted extraordinarily well to environments all over the world. Dioramas with live specimens and models show how insects and other arthropods have made themselves at home in four very different habitats.

In the desert diorama, you can see arthropods that have adapted to the special challenges of heat, water shortages, and evaporation. Darkling beetles get along on the tiny amount of water they extract from seeds, while centipedes wait until the cool of night to hunt.

The pond diorama features insects in a variety of microhabitats. Water striders skate along the surface, while dragonfly nymphs burrow in the mud.

Arthropods of the mangrove swamp have evolved ways to cope with constant changes in water level, chemistry, and temperature. Land crabs and termites stay well concealed. Mangrove and fiddler crabs slow down their breathing when temperatures rise. Shrimps regulate the seawater concentration in their bodies.

The tropical rainforest is home to millions of insect species. You'll find giant cockroaches on the forest floor, heliconia butterflies laying eggs on passionflower leaves, and insects and spiders living in bromeliad plants.

With their unsurpassed ability to adapt, insects will probably always be around to fascinate and annoy us.

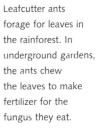

Leafcutter ants forage for leaves in the rainforest. In underground gardens, the ants chew the leaves to make fertilizer for the fungus they eat.

SPOTLIGHT SMITHSONIAN BUTTERFLY GARDEN

On the east side of the Museum building, a beautiful butterfly garden provides a peaceful oasis as well as information on the intimate connections between butterflies and the plants they feed on.

Depending on the time of year, you may see Monarchs, Viceroys, Tiger Swallowtails, Silver-spotted Skippers, and many of the seventy-five other butterfly species that have been identified within the District of Columbia.

The garden features four habitats frequented by butterflies—a wetland, meadow, wood's edge, and typical backyard. Interpretive labels illustrate frequent butterfly visitors and provide observation tips. Follow a butterfly for a few minutes. Stand quietly, and let butterflies land on you. Look for butterflies perching, feeding, and warming themselves.

Above: Monarchs drink nectar from many different plants but lay their eggs only on milkweeds.

In warmer months, the rich plantings, winding paths, and comfortable benches of the Museum's butterfly garden provide an inviting oasis.

BONES

Meet the vertebrates—fishes, amphibians, reptiles, birds, and mammals. These animals have an internal skeleton with a backbone consisting of a row of bones called vertebrae, which gives the group its name. As different kinds of vertebrates adapted to new habitats, seemingly endless variations on the basic skeletal structure evolved.

This Museum oversees a vast collection of vertebrate skeletons. Many highlights of the collection are on display in this hall. Museum scientists study bones like these to learn about the evolution and diversity of vertebrates

FISHES

The first internal skeletons evolved in primitive fishes 400 million years ago. Two skates are examples of cartilaginous fishes. The large, undulating fins that propel them through the ocean are supported by cartilage, not bone. You can compare their skeletons with those of bony fishes like the huge swordfish and the large blue catfish nearby.

The high dorsal fin of this huge swordfish helps control rolling when swimming in the sea.

PERCIFORM FISHES (Order Perciformes)

This order includes many species of advanced types of spiny-rayed fishes. It is a "catch-basket" order for many relatively unspecialized forms which were probably ancestral to most of the more specialized orders.

Like most advanced types, perciform fishes have spines in the fins, pelvic fins placed on the chest, and the maxillary bone excluded from the biting part of the upper jaw. Otherwise the group is best characterized by its lack of specializations found in other groups.

AMPHIBIANS

Right: The largest living turtle, the leatherback lives in the sea and can attain lengths of up to 2.4 m (8 ft.).

One feature you won't find on the fish skeletons is limbs. Limbs evolved with the first vertebrates to conquer the land—amphibians. Look closely at the frog skeletons to see how short their ribs are compared with those of fish, and how long the legs have grown in hopping species.

REPTILES

Among the most impressive reptile bones are the sinuous skeletons of a boa constrictor, Indian python, Eastern diamondback rattlesnake, and dozens of smaller snakes. Crocodilians are the largest group of modern reptiles. You can compare the broad snouts of the American alligator and American crocodile with the extremely long and narrow snout of the gavial. The turtles also invite interesting comparisons. The shell of the land-living Galápagos tortoise is high and domed. But in the two sea turtles—the green turtle and the leatherback turtle—the body and shell have become flattened to facilitate rapid movement in water, and the limbs have evolved into large paddles.

This frog's long hind legs unfold and extend to propel it forward in enormous leaps.

BIRDS

With the evolution of birds, vertebrates took to the air, leading to a whole new set of skeletal adaptations. You'll notice significant differences between the skeletons of birds that fly and those that don't. Consider the ostrich, for example. Its skeleton is large and heavy, with long leg bones for running on land. The frigatebird, on the other hand, is adapted for an aerial lifestyle. Although it has a wingspan of 2.1 m (7 ft.), its skeleton weighs less than its feathers!

A case of bird skulls illustrates how bills and beaks evolved for collecting and capturing different kinds of foods. The flamingo has a large, curved bill with serrated edges for straining tiny invertebrates out of mud and water. The parrot, on the other hand, has a sharply hooked beak that not only cracks open nuts and fruits, but also helps the bird climb.

MAMMALS

Mammals were the last vertebrates to evolve, and you can compare skeletons from the three major mammal groups. The echidna and duck-billed platypus are the only mammals that lay eggs and the only living members of the most primitive mammal group, the monotremes. The pouched mammals, or marsupials, are represented by a kangaroo and koala from Australia as well a Virginia opossum.

The diversity within the largest living group of mammals, the placentals, is impressive. Among the hoofed mammals are an Indian rhino, a zebra, a dromedary camel, a bison, and a completely articulated giraffe. Suspended above is a huge gray whale. Whales evolved from land mammals. As they adapted to the sea, their limbs evolved into fins.

Above: The tern is a consummate flyer. All the muscles needed for flying are attached to its large, keeled breastbone.

Like most snakes, the pit viper has vertebrae that seem to go on forever.

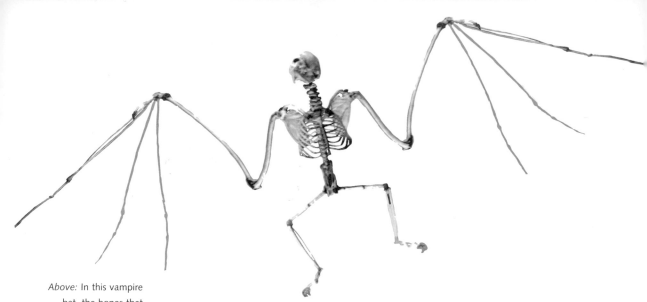

Above: In this vampire bat, the bones that make up the fingers have become modified to support wings.

Come face-to-face with the bones of primates like yourself, the gorillas, the chimpanzee, and the orangutan.

The teeth and claws of the cats, bears, dogs, seals, and other carnivores are specialized for catching and killing other animals. Look up to see one of the collection highlights: a Stellar's sea cow, which is more closely related to elephants than to seals and which is now extinct. The massive skeleton is a composite made of bones salvaged on the beaches of Bering Island in 1883.

A caseful of monkeys highlights adaptations for climbing: long arms and legs, tails for balancing, and hands and feet capable of grasping. The ape case reveals that the human ape has evolved for walking, with long legs and an erect posture. Our close relative the orangutan, on the other hand, has relatively short legs but very long arms for moving through the trees.

GALÁPAGOS

The Galápagos Islands are legendary in the history of biology, because their unique wildlife contributed to Charles Darwin's theory of evolution.

What intrigued Darwin was how the plants and animals had adapted to island life. Those species that survived on the Galápagos passed on their adaptations to their offspring. Over time, this process of natural selection resulted in animals and plants with different characteristics from their mainland relatives.

Isolated geographically and no longer able to reproduce with their ancestral populations, many animals and plants eventually evolved into new species. For example, the lack of large grazers on the islands may have presented an opportunity for other species such as the giant tortoise to flourish. The reason the island's cormorants can no longer fly could be the absence of large land predators. Also note the uniquely shaped beaks of three finches that are not found anywhere else.

Search for this case in the *Reptiles and Amphibians* exhibit outside the *O. Orkin Insect Zoo* to find examples of animals that now thrive in the Galápagos.

The Galápagos case features plants and animals from the islands today, including some unusual marine species collected by Smithsonian scientists.

Scientists estimate that there are more than 5,000 species of amphibians and 7,200 species of reptiles. Both of these highly successful groups far outnumber mammals.

Amphibians and reptiles are often grouped together because both are cold-blooded, or ectothermic—meaning that their internal body temperature is controlled largely by their environment. But reptiles and amphibians are very different from each other. Amphibians have smooth, moist skin; reptiles have scaly skin. Amphibians lay gelatinous eggs in water or moist surroundings, and their larvae change form dramatically as they grow into adults. Reptile young hatch from soft-shelled eggs laid on land or are born live, and they resemble miniature adults.

A central display introduces representatives of some of the major groups of amphibians and reptiles. The Gila monster is one of only two species of venomous lizards and an endangered species—despite its built-in defense. The gopher tortoise, which digs burrows in Florida sand ridges, is also threatened due to human encroachment on its habitat.

A diorama re-creates the Florida Everglades, where an American alligator presides over a variety of snakes, all looking for food.

You can see some of the biggest reptiles in the "Giants" case, including the world's largest lizard, the Komodo dragon of Indonesia. It grows up to 3 m (9.8 ft.) long. The dragon depends on its strong jaw, claws, and extremely sharp teeth to smash prey to the ground and then tear it apart. The largest venomous snake, Asia's king cobra, reaches lengths of 4 m (13.1 ft.) and is displayed here with its head raised and hood open.

Despite its name, the marine toad does not live in the sea. It has the largest geographic range of any amphibian.

Want to know how reptiles and amphibians protect themselves? Visit the case on defensive devices. Only a few snakes are venomous, and even fewer are lethal to humans. Curious about how they move around? Explore the locomotion case, where you may be surprised to find that reptiles and amphibians climb, hop, swim, crawl, burrow, walk, and run—every form of locomotion except flying.

Komodo dragons live on several Indonesian islands where they climb trees, swim, and chase down prey.

ANTHROPOLOGY HALLS

This complex of halls explores the human condition in all its diversity, and the forces that have influenced human biological and cultural development over time and around the world.

As members of the same species, all human beings are related. Yet differences show up in our cultures, languages, tools, and art. Located on the first and second floors, the anthropology exhibit halls display the everyday and the exceptional, the ancient and the contemporary. Over the coming years, the Museum will be updating older displays, opening temporary galleries that highlight new collections and Museum research, and developing new permanent halls in collaboration with members of the cultures that are represented.

Above: The famous brass heads made in Benin City, Nigeria, commemorate the region's rulers, called *obas.*

This face is part of the totem "Long Sharp-Nose," which tells a story belonging to the Wolf clan of British Columbia's Tsimshian people.

DAVID H. KOCH HALL OF HUMAN ORIGINS: WHAT DOES IT MEAN TO BE HUMAN?

Who are we? Who were our ancestors? When did they live?

The Museum's groundbreaking new *David H. Koch Hall of Human Origins*, scheduled to open in late 2009, explores these universal questions, showing how the characteristics that make us human evolved against a backdrop of dramatic climate change. The story begins 6 million years ago on the African continent where the earliest humans took the first steps toward walking upright. Early humans eventually spread to Asia and Europe as they adapted to new environments and climates. There have been over a dozen species of early humans, with multiple species often sharing the Earth. All of these species are now extinct—except for our own, *Homo sapiens.*

This incredible story is told through more than 280 fossils, casts, and artifacts, many of which are from the Museum's own collections. Presented alongside research from the Smithsonian Human Origins program and other scientific institutions, these objects illuminate the evolutionary history of our small branch of the tree of life.

A troop of *Australopithecus afarensis* moving through an ancient East African forest by artist John Gurche.

A JOURNEY BACK IN TIME

To enter the hall visitors pass through an atmospheric tunnel and travel back in time. Nine early human species appear and disappear with background silhouettes that depict their evolving human traits. Images of different environments also come and go, conveying the intense climate fluctuations that accompanied human evolution.

In the introductory area at each entrance, visitors come face-to-face with the skulls of five early human species. Their different sizes and shapes highlight changes in the human face and braincase over millions of years, providing evidence that modern humans evolved from early humans. Nearby, a large family tree shows where these early human species fit into the four groups that make up the human branch of the Order Primates.

Within the primates, humans belong to the great apes, one of the major groups on the primate family tree. By decoding human DNA, scientists have discovered that chimpanzees are our closest relatives. An amazing story of adaptation and survival is etched in our genes—and in the fossils in this hall.

Male and female Neanderthals *(Homo neanderthalensis)* in an imagined group portrait by John Gurche.

BECOMING HUMAN

A rich mosaic of physical traits and behaviors defines us as human. These traits and behaviors did not evolve all at once. Becoming a modern human took millions of years.

Against one large wall of the hall, a dramatic display of fossils, objects, videos, and images features some of the most significant milestones in becoming human: walking upright, making tools, evolving different body types and larger brains, developing social networks, and creating symbols and language.

Visitors to this area learn to look at fossils with a scientist's eye. For example, one skeleton on display, nicknamed "Lucy," has long, curved fingers that ensured a secure grasp in the trees, while its thigh bones angle outward at the knee joint, keeping the knees directly under the middle of the body—a requirement for efficient walking. This combination of apelike and humanlike features marks early humans' transition to upright walking. The ability to walk as well as climb enabled

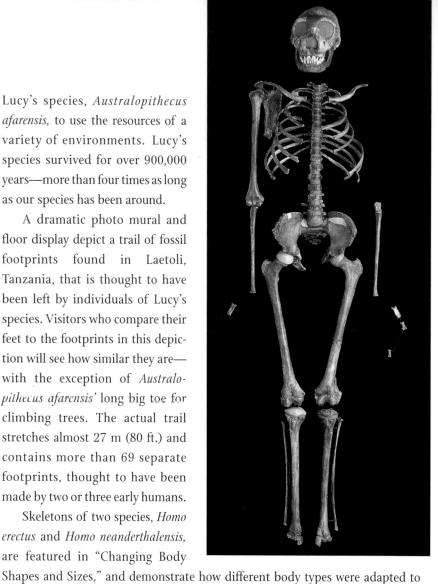

Lucy's species, *Australopithecus afarensis*, to use the resources of a variety of environments. Lucy's species survived for over 900,000 years—more than four times as long as our species has been around.

A dramatic photo mural and floor display depict a trail of fossil footprints found in Laetoli, Tanzania, that is thought to have been left by individuals of Lucy's species. Visitors who compare their feet to the footprints in this depiction will see how similar they are—with the exception of *Australopithecus afarensis'* long big toe for climbing trees. The actual trail stretches almost 27 m (80 ft.) and contains more than 69 separate footprints, thought to have been made by two or three early humans.

Skeleton of the Nariokotome Boy, a juvenile *Homo erectus*, found in 1.6-million-year-old sediments west of Lake Turkana, Kenya.

Skeletons of two species, *Homo erectus* and *Homo neanderthalensis*, are featured in "Changing Body Shapes and Sizes," and demonstrate how different body types were adapted to different environments. The *Homo erectus* skeleton, an 11- or 12-year-old male from Kenya, is 1.6 m (5 ft. 3 in.) tall. The long, narrow body and long legs—adaptations that help heat escape—were well suited to life in Africa where it was hot and dry. The *Homo neanderthalensis* skeleton, an adult from Europe, looks broad and short in comparison. The short, heavily built body retained heat and was well adapted to the cold winters and ice ages of Europe.

STORIES FROM THE FIELD

Opposite: At Olorgesailie, Kenya, a multinational crew excavates an elephant skeleton that was butchered by *Homo erectus* around 990,000 years ago.

Dramatic re-creations of three moments in time invite visitors to explore actual excavations. In Swartkrans, a South African cave discovered in 1948, researchers unearthed a 1.8-million-year-old youth with a leopard's fatal puncture marks still

Above: The cranium of OH 5, the famous *Paranthropus boisei* discovered by Mary Leakey at Olduvai Gorge in Tanzania.

A female *Homo erectus*, depicted in this bust by John Gurche, was one of the first human species to spread from Africa to Asia.

visible on its braincase. Visitors reconstruct the scene as they touch models of fossil "clues" from the site. After a narrator explains the significance of each fossil, the life-and-death events of that fateful day unfold in a time-lapse animation.

An elephant butchery site uncovered in Olorgesailie, Kenya, is the focus of another snapshot. Excavations at the site prove humans were advanced enough to use tools and work cooperatively nearly a million years ago. Finally, a 65,000-year-old Neanderthal grave in Iraq lined with flowers and tree boughs hints at the beginnings of ceremonial burial, a behavior never seen in earlier humans.

At the Olorgesailie Discovery Station, multimedia presentations feature Museum scientists working in the field and discussing fossils and ancient sediments that help them understand how the human species adapted to the tumult of the past 1.2 million years.

THE HUMAN FAMILY TREE

In a central display at the crossroads of the hall, an astonishing array of fossil skulls illustrates the history of human evolution. Highlights of this display include the skulls of *Sahelanthropus tchadensis,* the oldest known fossil human at about 7-6 million years old; *Paranthropus boisei,* or "Nutcracker Man," nicknamed for its big teeth and the strong chewing muscles that attach to a large crest on the skull; and *Paranthropus aethiopicus*, or "The Black Skull," which is the only known adult skull of this species. The skull's dark color comes from minerals absorbed from the soil as it fossilized.

So far, fossils of more than 6,000 individuals have been discovered, and more than a dozen species have been identified. To help make sense of this complex history, visitors can access several innovative computer programs to compare modern humans with their ancestors, find out how researchers distinguish one species from another, and see how theories and disagreements arise about how the various human species are related to each other.

Two views of *Australopithecus africanus,* STS 5, from Sterkfontein, South Africa, about 2.5 million years old.

MEET YOUR ANCESTORS

Lifelike faces of eight early human species look out from the central part of the hall. Artist John Gurche spent two-and-a-half years constructing the heads in this exhibit, using the latest forensic techniques, fossil discoveries, and his knowledge of ape and human anatomy to build up muscles, fat, and glands. All the heads are cast in lifelike silicone, with the skin tones based on typical pigmentation in the latitude where that species lived.

Starting with a replica of the fossil, Gurche uses his extensive knowledge of human and primate facial anatomy to build up muscles, fat, and glands. Shown here are stages in reconstructing the face of *Homo heidelbergensis* from Kabwc (Broken Hill), Zambia.

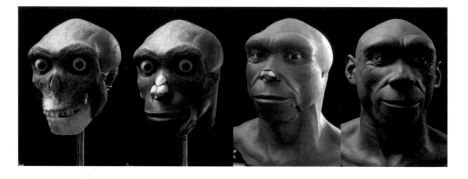

ONE SPECIES, LIVING WORLDWIDE

By 200,000 years ago, *Homo sapiens* had evolved in East Africa. From there—as a large world map of fossil discoveries shows—our species spread around the globe. Physical and cultural differences emerged as populations adapted to different environments. However, despite superficial variations in size, shape, skin, and eyes, all modern humans are remarkably similar, with DNA that differs by only 0.1 percent.

Artist John Gurche sculpted these two faces of *Australopithecus afarensis*. "Lucy" is on the left, a male of the species is on the right.

Left: This reconstruction of Shanidar 1, a male Neanderthal from Shanidar Cave, Iraq, was sculpted by John Gurche.

Below: The exhibit includes many reproductions of Paleolithic paintings and sculptures. This painting is of the yellow "Chinese Horse," from Lascaux Cave, France, about 17,000 years old.

KOREA GALLERY

Thousands of years ago, on a peninsula in East Asia, the distinctive culture and language of Korea arose. The *Korea Gallery* features several traditions that help define the nation's strong national identity, using artifacts from the Museum and other collections from around the world.

Elegantly presented in front of Korean latticework, a chronological display of ceramics, including classic Korean celadons, helps put Korea's long history in context, from its earliest kingdoms to the twentieth century. The wedding ceremony, another distinctive tradition, is illustrated with contemporary Korean wedding

Carved wooden *sotdae* are traditionally erected at village entrances to protect the community against calamities.

garments, designed by Korean designer Lee Young Hee and based on Joseon Dynasty clothing.

The practice of paying respect to one's elders, both living and dead, is based on Confucian principles that emphasize serving one's parents and ancestors. A family's ancestor altar shows how Koreans honor their family members today and in the past. For Korean families, celebrations of a person's first and 60th birthdays are especially significant.

Koreans are also very proud of their written script, called *hangeul*, which is based on the sounds of the spoken language. A kid-friendly interactive invites visitors to spell simple Korean words.

Just outside the gallery, contemporary Korean art illustrates that a dynamic, modern Korea finds inspiration in the rich traditions of its past.

Opposite Above:
Through the round hole on the side of this traditional *onggi* vessel, you can see a 3-D Korean cityscape designed by contemporary artist Y. David Chung.

Opposite Below:
A view of the Korea Gallery from one of the entrances.

AFRICAN VOICES

From every corner of this exhibition, African voices beckon. Proverbs call out their wisdom. Quotes from writers, historical figures, and everyday people interpret and authenticate the 400 objects on view. Developed through a close collaboration among Museum anthropologists, the local community, and African specialists, *African Voices* shows Africans making history, working, and creating art.

Three times the size of the United States, the continent of Africa encompasses fifty-four countries, more than 800 languages, and over 1,000 ethnic groups. An energetic media presentation at both entrances highlights this rich cultural diversity.

A historical pathway down the center of *African Voices* links six thematic galleries.

TAKE A WALK THROUGH TIME

Beginning with the origins of humankind and ending with contemporary challenges, this pathway highlights ten defining moments of Africa's history. Historical objects, maps, and time lines give each moment a sense of time, purpose, and place.

> If I stand tall, it is because I stand on the shoulders of many ancestors.
>
> **AKAN PROVERB**
> GHANA

On March 2, 1896, Ethiopians defeated invading Italians in the Battle of Adwa, depicted in this painting.

Below: A Moroccan *oudh* is on view in the *African Voices* Hall.

Casts of finely crafted, bone fishing tools, about 70,000 to 90,000 years old, represent Africa's ancient history. Three Egyptian *shabtis* from the seventh century BC—carvings placed in royal tombs to serve pharaohs in the afterlife—recall Africa's great Nile Valley civilizations.

Farther down the pathway, beautiful gold objects are displayed alongside eighteenth-century neck irons and shackles to illustrate this brutal chapter in African history. Gold drew Europeans to West Africa in the fifteenth century, but another trade became more profitable: slaves.

Near the end, contemporary objects document the overthrow of European colonial rule by one African country after another, culminating in South Africa's first multiracial democratic elections in 1994.

> **"** True worth is neither in the size nor in the weight, otherwise the ostrich would be a champion flyer among the birds.

TUKULOR PROVERB
SENEGAL

WEALTH IN AFRICA

Today, as in the past, African peoples create wealth through the exchange of goods and ideas. Wealth takes various forms—money, knowledge, and connections between people. For the Ga people of Ghana, well-attended funerals demonstrate the deceased's wealth in relationships and celebrate the journey to join the ancestors.

In some African societies, marriages are strengthened through gifts of bridewealth given by the groom's family to the bride's. Bridewealth honors all that the bride will bring to the new union—especially children. A tall iron blade, baskets, jewelry, and cash are among the forms of bridewealth on display.

The centuries-old tradition of Malian mud cloth, called *bògòlanfini*, is another example of African cultural wealth. Today, international designers of home furnishings and clothing have capitalized on mud cloth's strong graphic appeal. You can even admire contemporary mud-cloth fashions that have graced Paris runways.

Designed by Malian textile artist Nakunte Diarra *(above)*, this mud cloth, or *bògòlanfini, (below right)* incorporates traditional and new patterns.

This airplane coffin, designed and built by artist and master carpenter Paa Joe, proclaims the deceased family's prominence in Ghana's *Ga* community.

MARKET CROSSROADS

African markets bustle with the exchange of local, region-al, and global goods. This area re-creates several of the nearly 4,000 stalls that filled the downtown market of Accra, Ghana, in 1997. Life-sized cutouts introduce some of the respected market queens who supervise the market's day-to-day life.

> " The market is the world, the world is the market.
>
> **IGBO AND YORUBA PROVERB**
> NIGERIA

The kola vendors Adama Salifu and her two sisters have been selling kola for thirty years. Modern cola drinks originated from caffeine-rich extracts of the kola nut, a bitter-tasting nut with an honored place in West African cultures.

Cloth vendor Ernestina Quarcoopome and her daughter are part of a 100-ven-dor group that sells blue-and-white cloth with designs that draw on classic African proverbs and popular TV programs. Learn about the meaning of the cloths on display.

WORKING IN AFRICA

Innovation and tradition drive African workers. Africa has an 8,000-year-old history of agriculture, and farming remains the bedrock of most communities. But Africans also engage in many other kinds of work and celebrate its value through ceremony and art.

Nigerian potter Ladi Kwali learned her craft from her aunt and has become internationally famous.

For thousands of years, Afri-cans have made pottery. Most pot-ters are women, and pots—with their bellies and necks—are fre-quently compared to the female body. Compare the nineteenth-century pieces on display with the hand-built pots of contemporary Nigerian potter Ladi Kwali.

In contrast, most African met-alworkers are men, many of whom continue to produce traditional items like tools, emblems of wealth, and burial offerings. Other metal workers work in heavy industry or rework scrap metals into useful and beautiful objects.

Left: A few of the nearly 4,000 stalls in Accra, Ghana's downtown market, were re-created for the exhibit.

LIVING IN AFRICA

Stone houses and portable tents, public plazas and sacred palaces reflect diverse ways of life in Africa. Far more than shelter, these living spaces affirm the foundations of family and community and create connections among people and generations.

The Somali *aqal* is an example of a home designed for life on the move. Somali women construct these round homes from arched acacia roots and woven mats. The homes can be disassembled and packed on a camel's back. In a short video, a local Somali man and woman recall growing up in an *aqal*.

Somali women constructed this *aqal* especially for the exhibit hall. Included are typical contents ranging from baskets to an assault rifle.

Impressive buildings of coral stone connect the island of Zanzibar to an ancient legacy of Swahili stone cities along East Africa's coast. Furnishings such as the ivory-inlaid "Chair of Power" are typically made from durable materials, echoing the permanence of urban life.

Today, one in three Africans lives in cities, which draw residents from around the continent and around the world. An interactive display compares cities in Côte d'Ivoire, South Africa, Uganda, and Tunisia.

> The family is thus center and source,
> the hearth that maintains the flame of life.

LÉOPOLD SEDER SENGHOR
1959

KONGO CROSSROADS

Kongo people in Central Africa consider life a process shared with the ancestors, spirits, and a Supreme God. Powerful symbols, like crossroads, enable Kongo people to open portals to the invisible world and call on their ancestors for guidance. Enslaved Kongo people carried their beliefs to the Americas, where respect for the ancestors sustained them through adversity.

Graves are one intimate point of contact with the dead, as represented by the male and female grave markers from the Congo displayed here. You can also see an altar object from Brazil that reflects the meeting of Kongo people and American Indians.

GLOBAL AFRICA

For thousands of years, Africans have dispersed around the globe—sometimes willingly, sometimes not—building a community known as the Diaspora. A large, fiberoptic map traces the journeys that carried Africans to every continent, beginning with Hannibal leading his North African army against Rome in 250 BC.

A changing display highlights different Diaspora communities, and the Freedom Theater offers two videos on the Atlantic slave trade. One focuses on ways enslaved people created community and resisted oppression. The other explores common themes between the Civil Rights movement in the United States and the antiapartheid movement in South Africa.

FOCUS GALLERY

This intimate gallery presents temporary exhibitions on diverse themes.

Past exhibits have focused on the work of contemporary African artists and the blend of tradition and innovation in modern Ethiopian painting. "Discover Africa" invited young visitors to explore a Moroccan mud house; try on African clothing; and check out discovery boxes on topics as diverse as African textiles, minerals, and insects.

> " So truly, here I am at the crossroads. Where I was born. . . .
>
> KIANGI, A KONGO MAN
> 1971, CONGO

This alter object by Brazilian artist José Adário dos Santos uses the *dikenga* design, a symbolic crossroad encompassed by a circle.

> " We're related—you and I, You from the West Indies, I from Kentucky.
>
> LANGSTON HUGHES
> AFRICAN AMERICAN POET, 1924

ORIGINS OF WESTERN CULTURES

The institutions, traditions, and ideals of North American cultures are deeply rooted in those of western Asia, northern Africa, and Europe. Life-like dioramas and nearly 2,500 objects survey the increasingly complex systems of government, language, and beliefs that evolved in those regions between 20,000 BC and 500 AD.

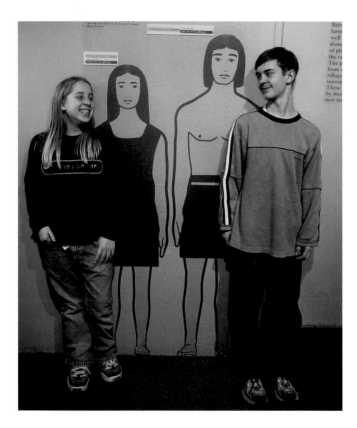

Throughout the *Origins of Western Cultures* hall, young and old can measure their heights against those of people from several ancient cultures.

FROM CAMP TO VILLAGE

About 10,000 years ago, nomadic people living across Asia and Europe began to see the advantages of herding over hunting and growing over gathering plants. Early archaeological evidence comes from Shanidar Cave in northern Iraq, where stone and bone tools show an increased diversity in size and shape. A diorama re-creates one of the earliest known farming communities—Ali Kosh, in Iran.

FROM VILLAGE TO CITY

Between 7000 and 3000 BC, farming as a way of life spread across Europe. As more kinds of plants and animals were domesticated, more specialized tools were created. Tools, weapons, and pottery from sites in Europe show this growing diversity. Many other trades also achieved higher levels of skill and innovation. By 5500 BC, metals were being heated and shaped. In one diorama, a metal smith from Larsa, in Mesopotamia, prepares bronze for casting into arrowheads.

About 3500 BC, cities began to emerge in southern Mesopotamia, supported by a growing population, an increased food supply, a system of irrigation, and a centralized government bureaucracy.

THE STATE

As urban life evolved, new forms of political and social organization arose. This area of the hall focuses on several advanced cultures: Mesopotamia, Egypt, and Troy.

Mesopotamia's need for records, communication, and trade produced cuneiform, one of the earliest forms of written language, about 4,500 years ago. On display are cones and seals that were marks of ownership and agreement.

Beginning about 2600 BC, the ability of the state to organize thousands of workers enabled the building of the famous Egyptian pyramids, temples, and tombs.

In this re-created Ice Age (30,000–10,000 BC) scene from Lascaux, France, two men paint a horse on a cave wall.

SPOTLIGHT THE ICEMAN

One of the most important archaeological finds of the twentieth century, the 5,350-year-old Iceman has revealed much about life in Europe more than 5,000 years ago. This new display includes a photomural about the Iceman's discovery in 1991 and a case with reproductions of his clothing and tools.

Genetic and microscopic analyses confirm that the Iceman was a European male about forty-five years old. Examinations of the body revealed cuts in his hand and an arrowhead in his back—two clues that confirm the Iceman's last moments were spent in a fight for his life.

All of the Iceman reconstructions—from his clothes and tools to the body and head shape—are based on actual remains and artifacts.

The Museum's displays of mummies and related funereal objects are especially popular. A male mummy, nicknamed "Minister Cox," is shown in its original linen wrapping and coffin. Much simpler mummification processes were used for lower class people and sacred animals.

"Treasures of Troy," the only Trojan collection in the United States, features 160 everyday items. The 2,000-year-old objects were donated to the Smithsonian by the wife of the nineteenth-century businessman and famous amateur archaeologist Heinrich Schliemann, who discovered the city of Troy in 1871 based on descriptions in Homer's *Iliad*.

TRADE AND EMPIRE

Trade united the ancient world and helped build many empires. By 1000 BC, the Phoenicians had probably established a trade route along the North African coast. A re-created dock at Tyre with timbers, an ivory tusk, and ingots of copper shows some of the raw materials traded here. A Pakistani bazaar includes stalls piled high with baskets, trunks, and textiles.

The Greek city-states competed with one another for new markets. Many of the nearly forty pots on display are like postcards from ancient Greece. They record historically important moments—athletic competitions, battles, and political events. Look through the large window to compare

Left: Mesopotamians recorded sales and taxes on clay tablets and cones, using wedge-shaped cuneiform writing.

This Egyptian carving of a gaunt man suggests a time of famine.

"Minister Cox" is nicknamed after the American who donated this mummy to the Museum.

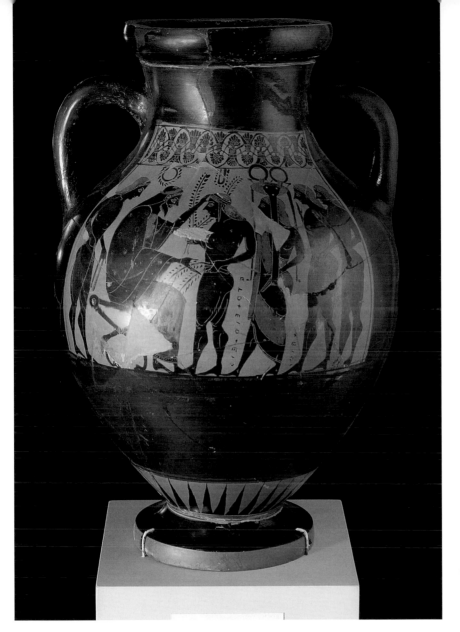

One of the Museum's most valuable Greek ceramics, *Crowning the Victor* depicts an athletic competition between city-states.

the architectural model of the Greek Acropolis with the classical columns of the Internal Revenue Service building—evidence of the continuing influence of ancient cultures on Western cultures.

Glassblowing was invented in Syria about 450 BC, but it was the Romans who perfected mass production and became a major exporter of glass. The broad influence of the Roman Empire is also evident in the distribution of iron farming tools and the expanding use of all types of literature and metal coins.

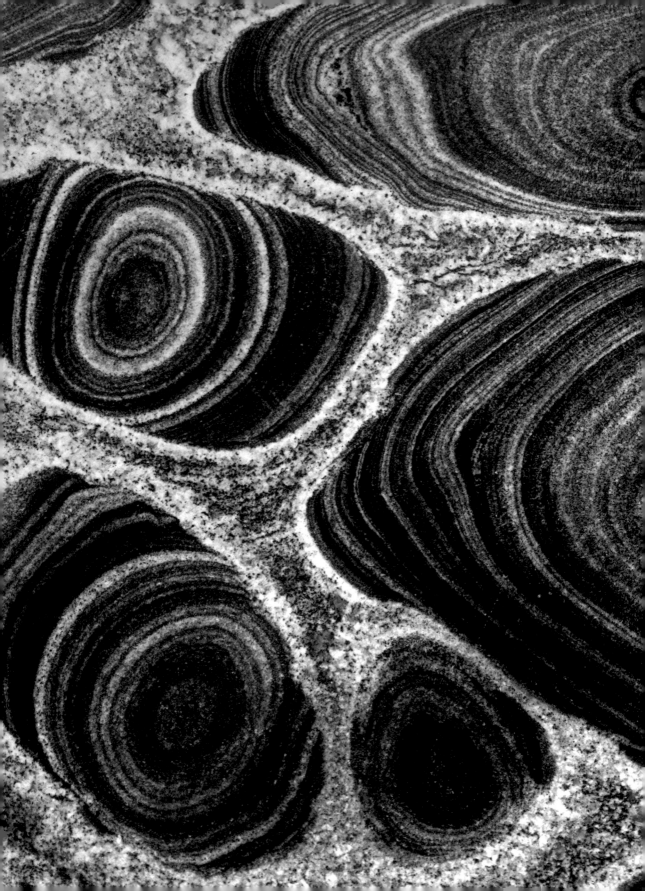

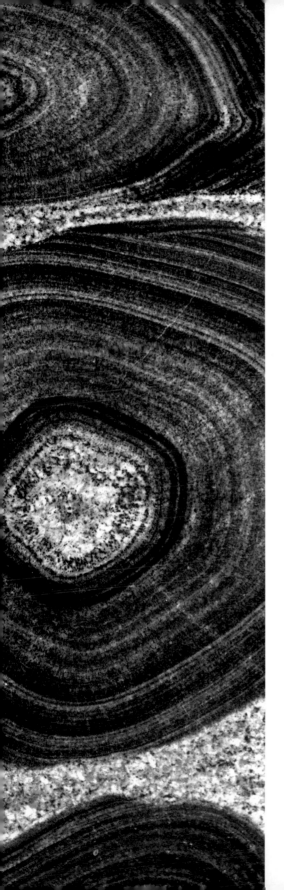

JANET ANNENBERG HOOKER HALL OF GEOLOGY, GEMS, AND MINERALS

From the fabled Hope Diamond to the meteorites that carry messages from other worlds, the specimens in this hall provide clues to the dynamic forces that formed—and are continually reforming—the Earth and our Solar System.

One of the largest renovations ever undertaken by the Smithsonian Institution, the Museum's 1,858 sq. m (20,000 sq. ft.) geology hall opened in 1997. Nearly 3,500 gems, minerals, rocks, and meteorites from the Museum's unparalleled collections are on display. Some highlights are featured in the first two galleries—the "Harry Winston Gallery" and the "National Gem Collection." Five thematic galleries follow, taking you from the tiny atoms that come together to form mineral crystals to the immensity of the Solar System.

This beautiful orbicular granite is on display in the "Rocks Gallery."

Above: This Martian meteorite is 1.3 billion years old. You can touch a piece of Mars in the "Moon, Meteorites, and Solar System Gallery."

The earliest fashioned gems in the "National Gem Collection," the large diamonds and emeralds in the Spanish Inquisition Necklace were probably cut in India in the seventeenth century.

The Hope Diamond's history of mystery and intrigue—as much as its size, beauty, and rare color—lure millions of visitors a year. The Hope was cut from the famous French Blue, a fabulous $112\frac{3}{16}$-carat diamond brought from India to France in 1668. King Louis XIV had the diamond recut, reducing it to $67\frac{1}{8}$ carats. More than a century later, during the French Revolution, the diamond mysteriously disappeared.

A $44\frac{1}{4}$-carat blue diamond—almost certainly the missing French Blue—turned up in London two decades later and was apparently sold to England's King George IV. From there, it came into the hands of Henry Philip Hope and then passed back and forth between diamond merchants and collectors in London, New York, Paris, and Washington, D.C. Finally, in November 1958, jeweler Harry Winston presented the Hope Diamond to the Smithsonian Institution and the American people.

Blue is one of the rarest diamond hues. The color comes from light interacting with traces of boron in the diamond's crystal structure.

HARRY WINSTON GALLERY

Six masterpieces of nature surround you in this elegant, introductory gallery. They speak of powerful forces —wind and water at Earth's surface, heat and pressure below—that have been constantly reshaping our planet since it formed.

Holding the place of honor is one of the world's most famous gemstones: the Hope Diamond. Renowned for its flawless clarity and rare, deep blue color, the 44.25-carat gemstone is set in an intricate diamond pendant and pirouettes slowly in a glass vault. It formed more than a billion years ago and more than 150 km (93 mi.) deep within the Earth. After being carried to the surface in an explosive eruption, it came into human hands less than four centuries ago. Since that time, it has crossed oceans and continents and passed from kings to commoners and, finally, to this Museum.

The five other specimens are also spectacular. A 590 kg (1,300 lb.) cluster of quartz crystals was found in a mine in Namibia, Africa. A long sheet of nearly pure Michigan copper, weighing 147 kg (324 lb.), is one of the few of this size ever recovered. The polished slab of multicolored gneiss from Sri Lanka is the product of 1.1 billion years of heat and pressure in the Earth's crust.

The Hope Diamond holds the spotlight under an elegant dome in the "Harry Winston Gallery."

The famous Tucson meteorite is really two fragments—weighing 283 kg (623 lb.) and 621.5 kg (1,370 lb.) respectively—of a single iron meteorite. After their discovery in Arizona in the late 1700s, they were used by blacksmiths as anvils until their true identity was discovered.

Together, these Earth treasures provide a tantalizing hint of what lies ahead in the next six galleries.

This sinuous sandstone concretion from France formed when sand grains from an ancient beach were cemented together.

Left: The composition and internal structure of the Tucson meteorite—as well as the ring's shape—are unique.

NATIONAL GEM COLLECTION

The Museum's National Gem Collection includes more than 10,000 specimens and is considered one of the world's finest. You can see some of the best and brightest specimens in this jewel box of a gallery. Rubies, emeralds, diamonds, and sapphires—more than 7,500 individual gemstones—fill the cases. Almost all were gifts from individuals to the Smithsonian Institution.

Many of the pieces in this gallery once belonged to royalty. You'll find gems formerly owned by a queen, an emperor, a duchess, a countess, and several maharajahs. The large, pear-shaped diamonds once rested in a pair of earrings worn by Marie Antoinette, the French queen who was guillotined in 1793 during the French Revolution.

Treasures from the Museum's renowned National Gem Collection sparkle in a room just off the Rotunda.

Most diamonds are naturally tinted pale yellow or brown. Colorless diamonds are rare and valuable, and fancy-colored diamonds are even more rare. You can see diamonds in many wondrous hues in the colored diamonds case. Nearby, you can compare the 253.7-carat uncut Oppenheimer Diamond with the 127.01-carat cut Portuguese Diamond, the largest in the collection.

For a different perspective, look through the large crystal ball in the middle of the room. The world's largest flawless quartz sphere, its spherical shape produces an upside-down image.

Once the property of Abdul Hamid II, sultan of the Ottoman Empire, the 75.47-carat Hooker Emerald exhibits exceptional green color.

Far Left: The National Gem Collection is noted for its colored diamonds. It is estimated that only one in every 100,000 diamonds is fancy colored.

Left: The extraordinary 330-carat Star of Asia is said to have belonged to India's maharajah of Jodhpur.

MINERALS AND GEMS GALLERY

Upon entering this gallery, you come face to face with an astonishing three-dimensional image of countless geometrically arranged red and blue balls. All salt crystals, including the tiny grains in your salt shaker at home, have this cubic arrangement of atoms. And although their atoms are arranged in different ways, all mineral crystals have a similar orderly, repeating pattern of atoms.

In the gallery beyond, the colors, shapes, and diversity of 2,500 minerals and gems beckon. They are organized into six major sections, based on important mineral properties.

SHAPE

In the "Minerals and Gems Gallery," treasure cases down the center of the room feature collection highlights.

The shapes of some crystals in this gallery are so surprising that it's hard to believe that they grew naturally within the Earth. You'll find blocky cubes, thin needles, long blades, and myriad other forms. What causes these different shapes? The answer lies inside. Different kinds of atoms arrange themselves in particular patterns, determining each crystal's basic shape. As crystals grow, differences in

temperature and chemical composition cause fascinating variations.

Take pyrite, for example. All pyrite crystals are built of iron and sulfur atoms linked in a cubic pattern. But different growth conditions create dozens of distinct shapes. Calcite, another mineral known for its variety of forms, has a basic threefold pattern of crystal faces with over 2,500 distinct variations—more than for any other mineral!

COLOR

Since the dawn of civilization, people have treasured minerals for their dazzling colors and varied hues. The color results from the way atoms interact with light. A brief video illustrates how certain atoms absorb some of the colors in visible light. You see the colors that are not absorbed. The great variety of atoms and atomic arrangements in minerals yields a virtual rainbow of colors.

Turquoise and malachite always come in shades of blue or green. Why? They all contain copper. When copper atoms combine with oxygen, the atoms absorb all colors except blue and green.

Other minerals may take on strikingly different hues if they contain trace impurities, inclusions, or defects in their atomic structure. Pure corundum, for example, is colorless. But when chromium atoms replace some of its aluminum atoms during growth, the result is a red ruby. Iron and titanium impurities produce a different gem: blue sapphire.

A special 3-D technology enables visitors to look into a model of a salt crystal. The red and blue balls represent sodium and chlorine atoms, enlarged almost 1.5 billion times.

AMAZING GEMS

The minerals in this area play tricks on your eyes with their shimmering surfaces, radiating stars, and flashes of fire. If you turn an opal from side to side, for example, the colors dance. An electron microscope photo shows why: Tiny spheres of silicon and oxygen, stacked up like oranges, scatter light. These special optical effects make opals, and many of the other minerals in this area, highly desirable as gems. Light reflecting off needlelike crystals or hollow tubes produce the slits in cat's eyes and multirayed stars. Among the stunning examples on display are a twelve-rayed star and the deep-blue Star of Bombay, given by Douglas Fairbanks, Sr., to his wife, silent film actress Mary Pickford.

DIVERSITY

Earth is a vast and varied chemical laboratory, with an enormous range of environments. Different temperatures and pressures act on the atoms of various elements to create the 4,000 minerals scientists have identified so far.

Some minerals, like the cavansite from India, are rare in the Earth's crust. Quartz is one of the most common, and a large case in the center of this gallery exhibits an astonishing variety of forms and colors. Also on display are three huge slabs of quartz crystal that came to the Museum in 1940 from a mine near Hot Springs, Arkansas. The largest slab weighs about 400 kg (880 lb.) and contains about a thousand crystals!

GROWTH

Crystals grow! Most grow from water rich in dissolved minerals. But crystals also grow from melted rock and vapor. The process starts when groups of atoms lock

Below Left: These pyrite crystals formed exactly as you see them. Because of its color and luster, pyrite is also known as fool's gold.

Below Right: Rhodochrosite is always pink or red because it contains manganese. In this extraordinary specimen, it is combined with quartz.

together in repeating, three-dimensional patterns. Then, under the influence of changing temperatures and pressures, wonderful and mysterious things can happen. You'll see a variety of examples: crystals that grew on top of each other; crystals within crystals; crystals etched by water; crystal twins; cup-shaped crystals; crystals too small to see without a magnifier; and many more amazing variations.

You'll also see crystals that grew inside pockets or cavities, including a group of giant gypsum crystals from Mexico's Cave of Swords.

PEGMATITES

Many of the largest crystals in the mineral kingdom come from pegmatites, exceptionally coarse-grained formations born of molten rock beneath Earth's surface. A 650 kg (1,430 lb.) touchable beryl crystal shows how large crystals in pegmatites can grow. You can also peek into the recreated pegmatite pocket.

Pegmatite deposits contain over 300 minerals, and some of them provide important industrial materials.

Above: The 318.4-carat Jubilee Opal is a black opal from a mine in Cooper Pedy, Australia.

Above: Smoky quartz grew first, followed by a rose quartz frill. This specimen is sometimes called the Pink Tutu.

Left: With its exceptional size, honey color, and sharp band of light, the 58.2-carat Maharini Cat's Eye from Sri Lanka is one of the finest gems of its kind.

Beryllium, the second lightest metal known, is scarce in the earth's crust but concentrated in beryl and other pegmatite minerals. It is an important component in high-strength alloys for rockets, satellites, and jet airplanes.

Pegmatites are also a source of spectacular gems. You won't want to miss the beautiful topaz, tourmaline, and other examples on display.

Left: This large topaz crystal, capped with lepidolite, came from a pegmatite.

Left: As these three crystals from California's Tourmaline Queen mine grew, the solution changed from manganese-rich to iron-rich, capping each hot-pink crystal with a band of blue.

Opposite: These huge, bladelike gypsum crystals were discovered by miners about 244 m (800 ft.) underground in the Cave of Swords in Mexico.

MINE GALLERY

Lights dim, rock walls rise up on all sides, and you are deep within the Earth, touring four remarkable mines from different parts of the United States. The veins and pockets were re-created using minerals and rocks from the actual mines.

Just inside the gallery entrance you'll find a case where everything that glitters really *is* gold. Along with platinum, silver, and copper, gold is a native metal—an element that occurs in nature chemically uncombined with other elements. Some of the largest pieces in the case—including an impressive triple ribbon weighing 455 g (1 lb.)—come from California's Mother Lode, site of the 1849 California Gold Rush.

This large gold nugget from California's Mother Lode weighs in at 2.55 kg (5.6 lb.)!

Below: In the "Mine Gallery" you can follow a winding corridor past four re-created mines.

The first diorama you'll encounter is the Fletcher mine, which is typical of lead-zinc mines in Missouri and neighboring states. A shaft drops 290 m (950 ft.) to a shallow, nearly horizontal vein of ore. The major ore is galena. Nearby, shimmering bands of blue-green amazonite light up a tunnel through Virginia's Morefield Mine. Morefield is mined for amazonite—an ornamental stone found only in pegmatites.

The classic mining town of Bisbee, Arizona, is famous worldwide for its diverse and magnificent minerals. It has produced more than 200 kinds—most notably azurite, malachite, and other copper minerals. In the diorama, a mine tunnel breaks into a natural underground cave in the famous Copper Queen mine, revealing sparkling pockets of azurite and malachite.

Ultraviolet lights enable you to see a mine that literally glows in the dark: Sterling Hill, New Jersey. More than seventy minerals found at Sterling Hill and nearby Franklin fluoresce, creating a riot of color. Smithsonian geologists have been studying this unique mineral deposit, once mined for zinc, since the 1930s.

Above: This re-creation of an underground cave in the Copper Queen mine, Bisbee, Arizona, includes stalactites colored blue and green by copper.

Under ultraviolet light, the ores In this deposit from New Jersey's Sterling Hill mine blaze with color.

ROCKS GALLERY

At Earth's surface and deep below, rocks have been continually forming and changing from one kind to another since our planet's birth 4.6 billion years ago. Each rock in this gallery preserves a bit of Earth's history.

A mosaic of lava columns, collected in Washington State for this exhibition, holds center stage and illustrates how rocks record the events that

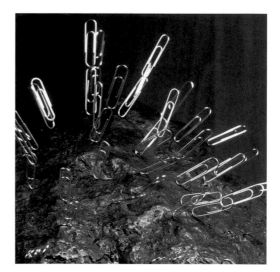

created them. The columns originated in Earth's interior as molten rock that erupted 15.7 million years ago and shrank as it cooled, opening up vertical cracks. As the cracks deepened, columns as high as 10 m (33 ft.) formed.

In "Rock Snapshots," you can pair six rocks with the fleeting moments that they preserve. Which was tossed out of a volcano? Which was hit by lightning? Nearby, three large rocks tell epic stories that span millions of years. A computer interactive shows how each formed. In the microscope area, you can see closeup how minerals come together in different ways to create different kinds of rocks.

ROCKS BUILD CITIES

Throughout history, cities have been built with rock—and Washington, D.C., is no exception. Through the large window in this gallery, you can see the U.S. Capitol, built in part from the same 100- to 120-million-year-old sandstone that was quarried to build the White House beginning around 1790. The sandstone is attractive and easy to carve, but

Above: This sandstone baluster once adorned the East Front of the U.S. Capitol. Parts weathered away, and it had to be replaced.

Like other sedimentary rocks, this beautiful sandstone formed as layers of sediments hardened.

not very durable. On display are several carved brackets and a baluster from the Capitol, heavily painted to protect the easily weathered stone. You can compare them with the more durable slate and granite used to build this Museum.

WATER RECYCLES ROCKS

Water is a great breaker, mover, and maker of rock. Rain, snow, ice, and running water demolish even the hardest rocks—then build new ones from the rubble—in an endless cycle. Rocks that form in this way are called sedimentary, and dozens of striking examples are on display. For example, you can see a geyserite that formed when steam and hot water erupted at Yellowstone National Park, or touch a piece of limestone with grooves left by a glacier 10,000 years ago. A walk-in cave diorama enables visitors to experience the kind of subterranean wonderland that results from acidic groundwater percolating through limestone.

ROCKS MELT

Deep within the Earth, it is extremely hot—hot enough to melt rock. With every kilometer in depth, the temperature increases by about 25°C (45°F). On display are a variety of rocks that originated deep within Earth as hot, molten lava. They

This beautiful granite has egg-shaped orbs of minerals that formed under different temperatures and pressures.

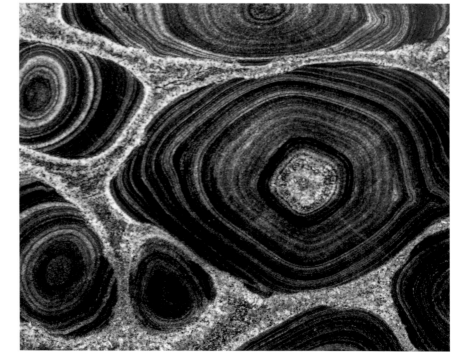

This schist includes large crystals of the mineral kyanite, a clue that it metamorphosed within the Earth at high pressures and moderate-to-high temperatures.

are called igneous rocks, from a Latin word meaning "fiery." Igneous rocks that took shape slowly in Earth's warm interior are called plutonic. Those that were spewed from Earth's interior to cool on the surface are called volcanic. An interactive explains how to tell plutonic and volcanic rocks apart, and challenges you to identify four mystery rocks.

ROCKS TRANSFORM

Like a caterpillar that changes into a butterfly in the quiet of its chrysalis, a rock's minerals can change from one set to another under the incredible temperatures and pressures deep within the Earth. This process is called metamorphism, from a Greek word that means "to transform."

Metamorphism on a grand scale occurred in New England 420 million years ago, when two continents collided. North America's eastern margin was shoved deep beneath Earth's surface. In the hotter surroundings, limestone turned to marble; basalt to amphibolite or granulite; and shale to phyllite, schist, or gneiss. A large display case shows how geologists have been able to reconstruct part of the history of the Appalachian Mountains by mapping the locations of these rocks.

PLATE TECTONICS GALLERY

Some 200 million years ago, a supercontinent split apart, and the Atlantic Ocean was born. Last year, the Atlantic widened by about 2.5 cm (1 in.). In another 200 million years, it may be twice as wide. Slowly, but relentlessly, propelled by intense heat beneath the crust, Earth's surface has been reshaping itself since the planet's birth 4.6 billion years ago. This process, called plate tectonics, is still happening today. A short film presented in the Plate Tectonics Theater shows this dramatic process at work.

The centerpiece of the gallery is a giant globe that highlights the mosaic of eight large rocky plates and nine smaller ones that make up Earth's surface. These huge slabs of rigid rock ride on a layer of hot mantle rock and move imperceptibly as heat escapes from the planet's hot interior to cold outer space. A computer interactive explores how, over the past 200 million years, plate tectonics broke up the supercontinent of Pangaea and rafted the pieces to form the seven continents we know today.

A large globe in the center of the "Plate Tectonics Gallery" highlights the rocky plates that make up Earth's surface.

At plate boundaries, some of Earth's most dramatic geologic events take place. A large video monitor plots earthquakes and volcanic eruptions since 1980. Watch closely: As these events appear in sequence, the plates are defined.

VOLCANOES

Vivid reminders of the superheated magma flowing beneath Earth's crust, volcanic eruptions are driven by the same heat engine that powers plate tectonics. Volcanoes vary widely in size, shape, and explosivity. Compare the five volcano profiles and the products that erupted from them to compare the differences. The Hawaiian volcano Mauna Loa, for example—the largest on Earth—is a shield volcano, with broad slopes built up by repeated eruptions of fluid lava. A spindle bomb from the cinder cone Parícutin in Mexico testifies to a more violent birth. In these short-lived but spectacular volcanoes, clots of incandescent magma shoot into the air and harden into contorted shapes, while falling particles build a small, steep cone.

If your interest in volcanoes has been piqued, you can access the latest data from the Museum's highly regarded Global Volcanism Program at two computer terminals.

HOT SPOTS

What do the Hawaiian Islands and Yellowstone National Park have in common? They are both above hot spots—places where a plume of hot, solid rock rises through Earth's mantle, partially melts, and generates magma.

In Hawaii, a chain of volcanic islands formed as a tectonic plate moved over a hot spot. One of Earth's most active volcanoes, Kilauea, is currently sitting on the hot spot, and you can see fascinating products of recent eruptions. In Yellowstone, the hot spot powers the complicated plumbing system that produces the park's hot springs and geysers. Three times in the past 2.1 million years, the magma has

An interactive model of a typical, active volcano invites you to explore its anatomy.

erupted explosively, leaving huge calderas that you drive through when you visit Yellowstone. On display are specimens from each of these historic eruptions.

WHERE PLATES PASS BY

Why do earthquakes shake California? The state straddles two plates that are moving past each other like trains on opposite tracks. A zone of active faults, the most famous of which is the 1,200 km (750 mi.) long San Andreas Fault, marks the boundary between plates.

The San Andreas Fault is a transform fault, a kind common on the sea floor but rarely found on land. The highlight of this display is a slice from a trench dug across one of the many smaller faults in the San Andreas Fault system. It was excavated by Museum technicians in 1995. Interpretive labels help you read the geologic history recorded in the sediments.

The adjacent Earthquake Study Station provides more information about earthquakes. A live-time video display plots earthquakes on a world map as they occur. What happened today?

Hot and sticky when it erupts, pahoehoe lava hardens into a twisty, rumpled shape. This specimen comes from Hawaii's Kilauea Volcano.

In the Earthquake Study Station, visitors jump up and down to make an earthquake and watch the vibrations on a seismometer.

WHERE PLATES MOVE APART

Hidden beneath the seas, along a 65,000 km (40,000 mi.) long global system of mountain ridges, Earth's plates are growing and spreading apart. Each year these oceanic spreading ridges emit more than three times as much molten rock as do all the volcanoes on land. These eruptions have had a dramatic effect on the landscape of the northwestern United States and Iceland.

Just 400 km (250 mi.) off the coast of Oregon and Washington, the Juan de Fuca Ridge is spreading at a rate of 6 cm (2.4 in.) a year. A seafloor chimney, or black smoker, collected by scientists aboard the tiny research submersible *Alvin*, was built up through eruptions at these plate boundaries.

On the other side of North America, the Mid-Atlantic Ridge bisects the Atlantic Ocean for 11,300 km (7,000 mi.). It is hidden from our view—except at one spot where it straddles a hot spot. Volcanic activity from the ridge and hot spot have produced the world's largest volcanic island: Iceland. In 1783–84, the Laki lava flow—the largest in history—produced enough lava to cover Washington, D.C., to a depth of 85 m (279 ft.). A piece of lava from this enormous eruption is on display.

THE ANCIENT CONTINENTS

Want to avoid an earthquake? Then head for a craton. These vast areas of rock, which make up large parts of each of the seven continents, have remained intact for billions of years. From them, scientists learn about ancient tectonic events and how the continents came to be.

The North American saga began over 4 billion years ago, when fragments of continental crust began coalescing into microcontinents. They later collided to form the heart of North America. Though sediment has blanketed much of this craton, portions are still visible in Michigan, Wisconsin, Minnesota, and much of Canada. Rock specimens tell the story of North America's assembly.

A background mural illustrates black smokers erupting along the Juan de Fuca Ridge, where they may grow as tall as fourteen stories.

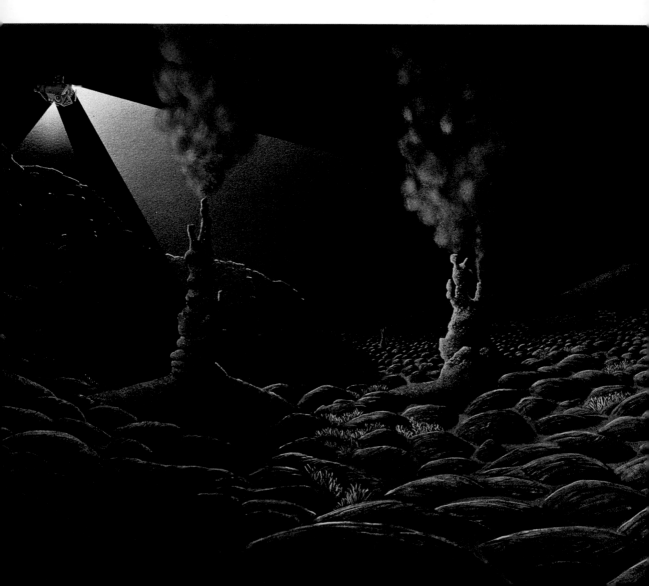

WHERE PLATES COME TOGETHER

Where plates come together, or converge, they are consumed and continents grow, and we see some of the most dramatic manifestations of plate tectonics.

In the Pacific Northwest, the snow-clad volcanoes of the Cascade Range sit above a subduction zone, where two oceanic plates are descending beneath the North American Plate, generating water-rich magmas that fuel explosive eruptions. Of the twenty-six active Cascade volcanoes, fourteen have erupted in the last 2,000 years—including Mount St. Helens in 1980.

In Asia, the dizzying peaks of the Himalaya provide some of Earth's most spectacular scenery. These towering mountain ranges were built when two plates collided, and they are still growing today at a rate of 1 cm (0.4 in.) a year.

Left: This tree trunk was splintered by the explosive eruption of Mount St. Helens in 1980.

Below Left: Preserved by the North American craton, this 3.96-billion-year-old gneiss from Canada's Northwest Territories is Earth's oldest-known rock.

Below Right: These tall pinnacles are made of pumice and ash from the colossal eruption of Mount Mazama in 5,700 BC—one of the largest in the last 10,000 years.

MOON, METEORITES, AND SOLAR SYSTEM GALLERY

Earth's story begins here, with the birth of the Solar System more than 4.6 billion years ago. Our planet—along with all the other planets, moons, asteroids, meteoroids, and comets—was a byproduct of the Sun's formation from a vast cloud of dust and gas. Few clues to these tumultuous events are left on Earth because plate tectonics and erosion have erased them. But meteorites—extraterrestrial rocks that land on Earth—preserve evidence of the Solar System's infancy and growth.

In this gallery's small theater, you can witness the birth of the Sun and all the other bodies in the Solar System. A short video at the other end of the gallery takes you on a tour of the Solar System today. Don't miss the chance to touch a piece of Mars in the small case nearby.

A mural shows how Arizona's Meteor Crater was blasted out by a 45 m (150 ft.) wide meteorite traveling more than 40,000 kph (25,000 mph) through space.

IMPACT!

Before there were planets circling the Sun, there were colliding grains of dust. Like rolling snowballs, large particles gathered up smaller ones. Some bodies grew huge, only to be destroyed by enormous impacts. Others survived. Over time, these encounters created the present array of planets and asteroids.

You can explore evidence of some of the collisions that left their imprint on the rocky bodies arrayed around the Sun and that continue to shape the Solar System today—including the enormous collision that created Arizona's Meteor Crater nearly 50,000 years ago. This enormous crater is the best-preserved impact structure on Earth, and the foremost natural laboratory for learning about impact cratering. Shattercones, tektites, impact breccias, and other specimens that result from impacts enable you to investigate what happened at Meteor Crater.

A computer interactive invites you to make your own impact by varying the size and speed of a meteorite. You can also watch a short video that shows how a huge meteorite wiped out the dinosaurs and changed the course of evolution 65 million years ago.

OUR CELESTIAL NEIGHBOR

Although the Moon bears little resemblance to the Earth, its origins are intimately linked with our planet's. Recent evidence suggests that a Mars-sized object collided with the infant Earth, vaporizing vast amounts of rock that went into orbit and condensed into a hot, partly molten Moon.

Four cases featuring rare rocks collected by *Apollo* astronauts tell the story of the Moon's history over the past 4 billion years. They describe how eras of bombardment and volcanism led to the relatively quiet sphere we know today.

This case houses a rare Moon rock. This display is one of a series that shows the different stages in the Moon's evolution.

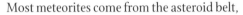

Right: The light-colored layer in this specimen provides evidence that a giant meteorite struck Earth 65 million years ago, wiping out 70 percent of all species—including the dinosaurs.

These touchable iron meteorites came from the cores of four asteroids that were shattered by cataclysmic collisions.

METEORITES

Messengers from space, meteorites provide an extraordinary opportunity to piece together the history of our Solar System's birth and evolution. Some are actual pieces of the original matter from which the planets were assembled. Others are fragments of asteroids that grew hot enough to melt and separate into core, mantle, and crust. Most are 4.5–4.6 billion years old.

Most meteorites come from the asteroid belt, located between the orbits of Jupiter and Mars. In 1970, Smithsonian scientists photographed a fireball over the central United States and traced its trajectory back to the asteroid belt. That historic meteorite, which landed in Lost City, Oklahoma, is on display here.

A wall of cases displays hundreds of meteorites from the Museum's collection. Learn how to distinguish a meteorite from a non-meteorite; examine meteorites called chondrites that preserve the Solar System's early grains; and see specimens collected in Antarctica, where more meteorites have been found than any other place on Earth.

Left: This small vial of stardust contains tiny diamonds forged in the explosion of a dying star and incorporated in the cloud that gave birth to our Sun.

The Gibeon meteorite's geometric structure offers hints about how iron meteorites form. The meteorite was found in Namibia in 1836.

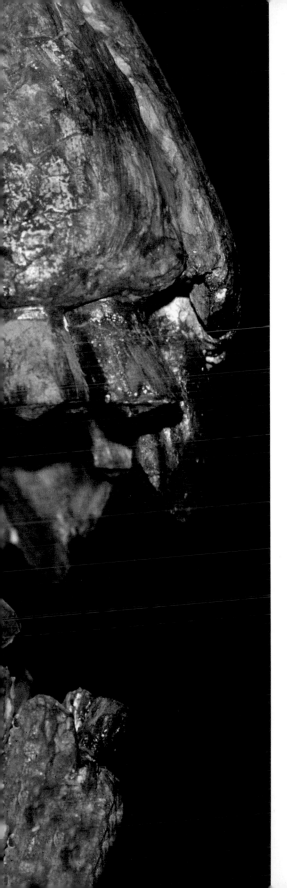

FOSSILS HALLS

From microscopic traces of early life to massive woolly mammoths, these halls tell the story of life on Earth—a story preserved in fossils.

Evidence of life's seemingly endless diversity fills the halls of the National Museum of Natural History. But how did life—and all its wondrous forms—come about? The story begins nearly 3.5 billion years ago, in the primordial seas that covered much of the Earth. Across the vast expanse of time between then and now, animals and plants have gradually evolved into the millions of forms that populate every habitat on Earth today. The more than 3,100 fossils in these halls tell this incredible story.

These fossilized teeth belong to *Camarasaurus,* a large dinosaur that lived 150 million years ago.

EARLIEST TRACES OF LIFE:
3.5 Billion–545 Million Years Ago

As the Earth and Solar System formed 4.6 billion years ago, the stage was set for the first steps of organic evolution. Volcanic gases in the Earth's atmosphere, along with energy from the Sun, spurred life's beginning. Highlighted in its own case is a fragment of the 4.6-billion-year-old Murchison meteorite, which is thought to be a remnant of our Solar System's birth. Deep within this ancient rock are traces of amino acids and other organic compounds, suggesting that meteorites may have carried some of life's essential ingredients to Earth.

Not until Earth had cooled enough to form a rocky crust, and water could condense and collect on its surface, could our planet support life. Several 3.8-billion-year-old rocks from western Greenland provide a rare glimpse of this young Earth. One is a small, unassuming cobble. Its rounded edges, tumbled and abraded by running water, prove that one of life's essential ingredients was present by this time.

ORIGIN OF LIFE

Although no one knows for certain when or how life began, scientists believe it arose from nonliving matter through a series of intermediate, chemical steps. Models and graphics explore conflicting theories about how life evolved.

One very special rock suggests that the crucial steps between nonliving and living matter may have been made as long as 3.5 billion years ago. Closely resembling sediments produced by

This view of
the fossils halls shows
the popular *Reptiles:
Masters of Land*
exhibit—home of
the dinosaurs.

microorganisms that form mats in today's shallow waters, the concentric layers in this rock fragment from western Australia may be the oldest-known direct evidence of life on Earth.

INTRODUCING THE MULTICELLULAR ANIMALS

From its early beginnings along the edges of shallow seas, life evolved from simple, single-celled forms such as bacteria and blue-green algae to more complex organisms. An animated film illustrates how multicellular life came about through trial and error over thousands of millions of years.

Rare fossils of soft-bodied organisms, taken from sandstone deposits in the Ediacara Hills of South Australia, introduce some of the animals that lived there about 555 million years ago. Several of these early animals resemble jellyfish found in the seas today. Others are unlike any known organisms and defy classification.

Above: A diorama offers a tantalizing glimpse of what life might have looked like in the shallow Ediacaran seas 555 million years ago.

Layers preserved in rock may be evidence of ancient stromatolites, structures formed by colonies of blue-green algae and bacteria. A mural shows how they may have dotted shallow waters 3.5 billion years ago.

GRAND OPENING—FOSSILS GALORE:
542–252 Million Years Ago

About 542 million years ago, complex animals with shells and hard skeletons suddenly appeared. This dramatic debut marks the grand opening of the Paleozoic Era.

Fossils Galore is a stunning display of Paleozoic diversity. Fossils of mollusks, sponges, corals, brachiopods, and arthropods fill the gallery cases, just as these invertebrates filled the shallow Paleozoic seas. No one knows for sure why so many complex, hard-shelled animals suddenly appeared, but Museum scientists are actively researching the question.

THE BURGESS SHALE— WINDOW INTO THE PAST

Nearly 505 million years ago in what is now Canada, shifting mud and silt from a submarine cliff buried living plants and animals in an oxygen- and scavenger-free environment. Even soft-bodied organisms were preserved in remarkable detail, providing a rare view into the past.

In 1909, Charles D. Walcott, the fourth secretary of the Smithsonian Institution, discovered this extraordinary fossil formation—known as the Burgess Shale—high in the Canadian Rockies. The Museum houses more than 65,000 Burgess Shale fossils, and some of the most significant from Walcott's rare collection surround a diorama that brings several of these bizarre animals to life. Many defy classification into any known group.

Some of the soft-bodied marine fossils preserved in the 505-million-year-old Burgess Shale of British Columbia are reconstructed in this diorama.

LIFE IN THE ANCIENT SEAS:
542 Million Years Ago to Today

Life began in the sea, and the history of life in the seas is a life-and-death drama on a grand scale. Twice, widespread extinctions have brought down the curtain on many kinds of marine life. Twice, another cast of actors evolved to fill the sea's stage. The marine fossils in this hall—together with lifelike models and spectacular murals—tell this dramatic story in three acts.

ACT I: THE PALEOZOIC ERA, 542–252 Million Years Ago

Life in the Ancient Seas spans 542 million years of marine-life evolution.

Fossils of flowerlike animals called crinoids, clamlike brachiopods, and other ancient invertebrates cover the walls of this gallery. An amazing assortment of trilobites, up to 71 cm (28 in.) in length, illustrates how these ancient arthropods had diversified. Nearby, fossilized fish specimens stand as proof that the first vertebrates had evolved.

By 250 million years ago, complex reefs were flourishing on the ocean floor. A colorful, full-scale reconstruction of a Permian reef gives visitors a rare glimpse into this ancient world. The result of forty years of marine-fossil research at the Museum, this dazzling reconstruction illustrates a marine metropolis teeming with a wide variety of organisms coadapted for crowded living. Over 100,000 models bring these ancient reef residents to life. You can even hear the sound of waves pounding against ancient shores.

The Paleozoic Era ended with the greatest mass extinction in the history of life. Nearly 90 percent of marine invertebrates went extinct, including the trilobites. Selected fossils, along with graphics illustrating the diversity of marine mammals before and after the extinction, chronicle this dramatic event.

ACT II: THE MESOZOIC ERA, 252–65.5 Million Years Ago

A new cast of characters emerges. Some, like fishes and mollusks, are descendents of survivors from Act I. Others, like marine reptiles and birds, are newcomers—animals that evolved on land and then took the plunge into the sea.

Delicate sea lilies, called crinoids, swarmed the shallow seas of the North American Midwest millions of years ago.

Along one wall, a dramatic floor-to-ceiling mural sets the stage. A giant sea turtle propels itself with large winglike forelimbs. Nearby, a flightless bird, *Hesperornis,* is poised to dive for fish. The remains of an ancient crocodile lie half buried on the reconstructed seabed below.

But the star of this show is a giant fossil mosasaur, over 9 m (30 ft.) long. These relatives of modern monitor lizards stalked shallow waters 70 million years ago, feeding on smaller mosasaurs, sea turtles, and squidlike cephalopods.

DIFFERENT STROKES

On a central island, an astonishing array of fish and ammonite fossils reveal how predators hunted in the Mesozoic sea. A 2 m (7 ft.) long relative of today's great white shark, *Squalicorax,* attacked prey with its triangular, bladelike teeth. Fossils of ray-finned fishes sport the physical adaptations that made them superb swimmers: thin, flexible scales; bony vertebrae; and symmetrical tail fins.

Nearly 100 fossil ammonites—cousins to today's nautilus—exhibit amazing shapes, sizes, and colors. Some glisten with iridescence. Others look like oddly-shaped corkscrews.

Once again, widespread extinction ended an era. On land, dinosaurs died out. In the seas, marine reptiles disappeared—along with many supporting actors like ammonites. The cause, this time, was an asteroid 10 km (6.2 mi.) in diameter.

ACT III: THE CENOZOIC ERA, 65.5 Million Years Ago to Today

The characters in this act are mostly survivors of Act II. Together with some notable newcomers, the marine mammals, they set in motion the marine drama that is still playing in oceans worldwide.

A long mural illustrates why this era is often called the Age of Mammals. Sea cows feed on thick underwater grasses, toothed whales plunge through cold waters, and seals dive for a fish dinner. In front, a fossil *Zygorhiza kochii* dominates the scene.

Suspended as if in midstroke, *Protostega gigas* was one of several huge sea turtles that swam in the shallow seas 80 to 70 million years ago.

This 39-million-year-old whale had a streamlined body with paddlelike limbs to maneuver in water, but it also had hind limbs like its presumed land ancestor.

In this same underwater setting, the fossil of a massive sea cow floats like a submersible through imaginary water. Carnivores also went to sea and, in the process, traded paws for paddles. Seals are among the featured carnivore fossils.

The teeth and elongated snout of *Zygorhiza kochii* (foreground) helped this early whale nab fish.

CONQUEST OF LAND:
420 Million Years Ago...the New Frontier

For billions of years, the land remained lifeless. Then, about 420 million years ago, plants began moving onto land, followed by animals. Conquering the land would take tens of millions of years, but it would reward the settlers with new sources of food and more room to diversify.

The "Scouts and Settlers" diorama illustrates what life on land may have looked like 400 million years ago. Scorpion-like animals crawl from a stream onto a shore carpeted with pioneer plants. You can examine a 400-million-year-old fossil of what might have been the land's first settler, *Cooksonia.* It is the earliest plant known to have some of the basic structural features that would have allowed it to live on land.

Fossil flowers are relatively rare compared to leaves, fruits, and seeds. They help scientists understand how ancient plants reproduced and were interrelated.

FISH OUT OF WATER

By adapting to survive seasonal droughts, fish laid the groundwork for other vertebrates to live on land. A diorama and fossil specimens demonstrate how two different groups of fish—lungfish and lobe-finned fish—coped in different ways when water dried up.

Lungfish burrowed into mud and became dormant until the rainy season restored fresh water. Lobe-finned fish crawled downstream on leglike fins to large, fresh pools. Then, 360 million years ago, lobe-finned fish crawled out of the water. Their descendants—the amphibians—continued the transition to living on land.

THE FIRST FORESTS

The first land plants took root in mostly sand and rock particles. But when they died, bacteria, fungi, and small animals fed on their remains, excreting organic matter and manufacturing the first soils. Growing in this rich humus, land plants grew bigger and more diverse. A forest scene takes you back 390 million years to see a small grouping of rare fossil plants and animals. Over the next 200 million years, descendants of these pioneer plants and animals would conquer the land.

The earliest land plants had no roots, cones, or flowers. They reproduced by spores, like ferns today. On display is 380-million-year-old *Rebuchia ovata*, complete with spore sacs. Then, about 350 million years ago, spore bearers gave rise to seed-bearing plants. Rare fossils of several early seeds and seed plants bear witness to this evolutionary milestone.

As plants evolved stems with more support, trees and forests appeared. In this gallery is the squat, fossilized stump of one of the earliest known trees, *Eospermatopteris*, which grew to an estimated 12.2 m (40 ft.). It's a relative of modern conifers and cycads, an ancient group of seed plants.

COAL SWAMP FORESTS

Primitive land plants reached their peak in numbers and variety during the Carboniferous Period, from 362 to 295 million years ago. Fossilized stumps, trunks, foliage, cones, and spores extracted from the great coal beds of Appalachia and other regions provide a glimpse of life in these ancient forests.

One of the major plants of the coal swamps, the scale tree reached heights of 38.1 m (125 ft.). Look for the spectacular fossil trunk of *Lepidodendron*, a huge scale tree from what is now Kentucky. On nearby fossil branches, leaf scars resembling fish scales explain why coal miners named these ancient plants scale trees.

A 4.9 m (16 ft.) tall fossil of *Callixylon*, one of the earliest trees, rises from the gallery floor. It is 348 million years old.

A table at the Fossil Café features fossil leaves of sycamores, a family of North American trees that has existed for 115 million years.

SEED PLANTS TRIUMPH

In time, the world's climate became drier, coal swamp forests shrank, and seed plants spread rapidly and diversified. One group, the gymnosperms, dominated forests for 200 million years. Named from the Greek words *gymno* (naked) and *sperma* (seed), gymnosperms have seeds without a protective cover.

The brilliant colors of a 215-million-year-old conifer trunk, *Araucarioxylon,* were caused by impurities in the minerals that replaced the woody tissue. Nearby, the fossil of a fan-shaped ginkgo leaf tells a surprising story. Smithsonian scientists have discovered that the number of tiny pores, or stomata, on these fossil leaves is directly related to the amount of carbon dioxide in the atmosphere, providing valuable information about past climatic change.

Walking through 3.6 billion years of evolution can be tiring. If you're thirsty or hungry, or need to rest, stop off at the Fossil Café, where you can explore fossil plants while you sip cappuccino. Seventeen tabletops double as fossil specimen cases. Explore the fossil evidence for plant-eating insects under small magnifiers, or examine the display of rare fossil flowers.

AMPHIBIANS—VERTEBRATES TAKE TO THE LAND

Around 350 million years ago, amphibians led the conquest of land by vertebrates. They laid their eggs in water but spent at least part of their adult lives on land, where they breathed air. A model of *Ichthyostega,* the earliest known amphibian, has stout, muscular legs adapted for walking. Its tail fin betrays its fish lineage.

Some amphibians were less successful than others at adapting to dry conditions, and their descendents went back to the water. Others became relatives to early reptiles. But for 50 million years, amphibians were the only vertebrates to inhabit the land.

REPTILES: MASTERS OF LAND

More than thirty of the Museum's 1,500 dinosaur specimens fill this soaring two-story gallery. Although some of the most delicate specimens have been replaced with models since the gallery opened in 1983, most remain the real thing.

Tyrannosaurus rex reigns over the entrance to Reptiles: Masters of Land.

An awesome predator and its potential prey, *Tyrannosaurus rex* and *Triceratops horridus*, face off at the gallery entrance. Beyond, a parade of colossal dinosaurs, some reaching lengths of 24 m (80 ft.), put on a show. But *Reptiles: Masters of Land* is more than the story of dinosaurs. It's the epic tale of the entire reptile family from its modest beginnings 310 million years ago. The development of the amniotic egg ended vertebrate dependence on aquatic environments and marked the revolutionary step from amphibians to reptiles. Free to occupy a much wider range of habitats, reptiles became masters of the land for the next 200 million years.

Life-size models, reconstructed from fossils, illustrate some of the earliest known reptiles. Small and slender, they look surprisingly like modern lizards. Yet these modest creatures eventually gave rise to two lines of reptiles: the synapsids, which led to mammals; and the sauropsids, from which dinosaurs and all other reptiles gradually evolved.

FIRST WAVE OF REPTILIAN DIVERSITY

More than 2.7 m (9 ft.) long and distinguished by its large dorsal fin, *Dimetrodon* is a primitive synapsid reptile that lived 280 to 250 million years ago. Its tall dorsal spines supported a membrane that may have functioned as a solar panel. Tiny *Thrinaxodon* lived nearly 230 million years ago. In the fossil, you can see its stabbing canine teeth, similar to those found in mammals.

A fossil skeleton of *Procaptorhinus* represents an early branch of reptiles that eventually gave rise to the dinosaurs, the huge reptiles of the Mesozoic Era, and modern reptiles.

EARLY DINOSAURS

Most of the earliest dinosaurs were graceful plant-eaters of moderate size. But a few reached the size of modern hippos. You can see models of several examples in a small diorama depicting a land community 220 million years ago. Two rhino-sized *Plateosaurus* wallow in the mud. The lizardlike *Trilophosaurus* scurries by as spiny-headed *Hypsognathus* strides off, leaving fresh tracks behind.

LIFE IN THE JURASSIC

By the Late Jurassic, 143 million years ago, dinosaurs were reaching sizes never before attained by land animals. In fact, Jurassic dinosaurs, and their Cretaceous descendants, were the largest animals ever to walk the Earth.

Dimetrodon grandis, an early reptile, gained heat by orienting its fin broadside to the Sun's rays, and lost heat by seeking shade.

Giant birds like Diatryma disappeared 37 million years ago because of mammal predators.

An impressive troop of Jurassic giants, from 200 to 142 million years ago, fills the center island of *Reptiles: Masters of Land.* At center stage is a 24.4 m (90 ft.) long *Diplodocus,* its long neck stretching high into the tall second level of the cathedral-like space. Discovered in Utah in 1923, the fossil took almost eight years to recover, transport, and mount. It is the largest fully mounted specimen in the hall.

Nearby, you can touch the 150-million-year-old upper arm bone of *Brachiosaurus altithorax.* And don't miss the baby duckbill dinosaur, *Maiasaura,* and a nest of eggs probably belonging to the small flesh-eating dinosaur, *Troodon.*

A newly cast *Stegosaurus* gives you a good idea of what this 5,080 kg (5 ton) herbivorous dinosaur looked like. This recent reconstruction corrects several misconceptions about *Stegosaurus,* including the number and position of its back plates.

CRETACEOUS DINOSAURS

Featured along the perimeter of the hall are dinosaurs that lived during the Cretaceous Period, about 142 to 65.5 million years ago. Many were vegetarians sporting ducklike snouts, but others were ultimate flesh-eaters.

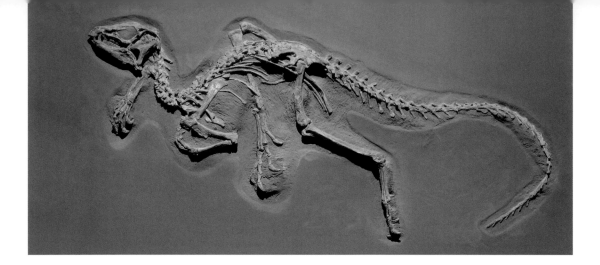

Chicken-sized *Heterodontosaurus tucki* belongs to a small group of dinosaurs that eventually gave rise to duckbills. In the remarkable mummified remains of the 4,064 kg (4 ton) duckbill, *Corythosaurus casuarius,* you can see rare skin impressions along the tailbone, indicating that this dinosaur had scales. The skull and teeth of *Edmontosaurus,* a large duckbilled dinosaur from Canada, illustrate how duckbills crushed food. Nearly sixty rows of teeth formed a broad shearing surface for mashing up tough plants.

Above:
Heterondontosaurus is one of the smallest, bird-hipped dinosaurs.

By the end of the Cretaceous Period, 65.5 million years ago, many dinosaurs reached mammoth proportions.

Tyrannosaurus rex, at 12 m (40 ft.) long and 4.9 m (16 ft.) tall, was one of the largest predators ever to walk the Earth. Dinosaurs ruled the planet longer than any other group of land animals. But about 65 million years ago, they went extinct with many other species when a huge asteroid hit the surface of the Earth.

You can feel feather impressions in the touchable cast of 145-million-year-old *Archaeopteryx,* the oldest-known bird.

SPOTLIGHT TRICERATOPS

The Museum's original *Triceratops,* on display since 1905, was a composite of seven different individuals. Today, only the head and lower jaw of the original *Triceratops* remain on exhibit. The newly mounted cast—named "Hatcher" after *Triceratops* discoverer John Bell Hatcher—has corrected feet, a 15 percent larger skull, and other improvements that make it more accurate.

When preparators dismantled the original *Triceratops* skeleton for conservation in 1998, they digitized each bone. Then, using computers, they analyzed how this extinct reptile walked 65 million years ago. A video shows this new *Triceratops* in the making.

In the current display, you can touch a model *Triceratops* horn or a reduced-size bronze cast of the *Triceratops* skull. Rare specimens and models show how *Triceratops* is related to other dinosaurs.

The *Triceratops* on display is the world's first totally digitized dinosaur.

SPOTLIGHT FOSSILAB

Paleontologist Steven Jabo prepares a cast molded from a fragile fossil in FossiLab.

Below: Visitors witness the painstaking process of molding casts from fragile fossils.

You never know what you'll find at FossiLab. It might be the latest dinosaur discovery, fresh from the field, being scraped from ancient rock before your eyes. Large glass windows allow you to watch Museum paleontologists and trained volunteers extracting fossils from rock or constructing fossil casts and molds. You can talk to them about their work. Or explore nearby exhibits to learn what a fossil is, why the Museum has so many, and how scientists use fossils to unravel the mysteries of evolution.

What's happening today?

MAMMALS IN THE LIMELIGHT

The earliest mammals appeared about the same time as the first dinosaurs, over 200 million years ago. But not until dinosaurs died out did mammals become stars of the show.

Mammals in the Limelight focuses on this spectacular explosion of mammalian life. Rare specimens and mounted skeletons—many of them unearthed in the American West by Smithsonian scientists—stand against a backdrop of vivid murals. Together, fossils and murals trace the remarkable evolution of mammals.

Greeting visitors to *Mammals in the Limelight* is a small skeleton of 50-million-year-old *Hyracotherium,* the oldest recognized ancestor of the horse.

EOCENE EPOCH: 55–33.7 MILLION YEARS AGO

By 55 million years ago, mammals had filled many of the niches left vacant by the dinosaurs. Gradually, forerunners of modern mammal groups replaced the primitive groups.

A remarkably clear picture of these mammals emerges from a treasure trove of fossils found in Utah, Colorado, and Wyoming. They include early members of many modern mammal groups, including horses, rhinos, bats, and rodents.

A dramatic mural depicts a subtropical forest in what is now the Rocky Mountain area of Wyoming and Utah. The giant skeleton of a hoofed mammal, *Uintatherium,* dominates the display. Despite its fierce-looking tusks, *Uintatherium* was a plant-eater. Another highlight is a primate—and one of your distant relatives. Lemur-like *Smilodectes* lived in Wyoming about 50 million years ago and was adapted for life in the trees.

OLIGOCENE EPOCH: 33.7–23.8 MILLION YEARS AGO

Cooling climates and receding forests increased the range of habitats for mammals—including squirrels, rabbits, peccaries, dogs, and beavers—and mammals adapted to life in a more open environment.

A second mural sets the scene in subtropical woodlands approximately 33 million years ago. Standing in front are two early rhinoceroses. *Trigonius* had a stocky body but lacked a nose horn. In contrast, *Hyracodon* had a slender, light build. They were among the largest herbivores of their time.

Mesohippus became the first horse with only three toes on each foot, enabling it to run faster than its four-toed relatives—an important step toward life in more open environments. A skeleton of an early camel relative, *Poebrotherium,* is about the size of a gazelle. Long slender limbs, with only two toes on each foot, made it the best runner of its time. And don't miss *Hesperocyon,* the oldest recognized member of the dog family.

MIOCENE EPOCH: 23.8–5.3 MILLION YEARS AGO

Grasslands spread as the climate became yet cooler and drier. To defend themselves and find food, Miocene mammals adapted to eating grass and moving quickly.

A third mural portrays North American grasslands about 20 million years ago. A horse-sized skeleton of *Moropus,* a chalichothere that emigrated from Asia, stands in front. *Chalichotheres* are related to horses, rhinos, and tapirs, but they had clawed feet instead of hooves. *Moropus* probably stood on its stout hind limbs and pulled leafy branches toward its mouth with long-clawed front legs.

Miocene herbivores developed tough teeth and long limbs. The camel *Stenomylus* was a swift runner with the acute vision needed to avoid predators on the open plains. At the same time, predators like the 8-million-year-old cat *Barbourofelis* evolved adaptations for pursuing and capturing fleet-footed herbivores. Its massive forelimbs and long saber teeth enabled it to prey upon any of the mammals of its day.

Horses reached their maximum abundance and diversity during the Miocene. Dramatic fossils and an animated film are featured in an alcove dedicated to the evolution of horses.

This nearly complete skeleton of lemur-like *Smilodectes* is one of the best specimens of Ecocene primates known.

LATE MIOCENE AND PLIOCENE EPOCHS: 10–1.8 MILLION YEARS AGO

By now, much of North America was savanna. Vast herds of herbivores roamed the grasslands, and intermittent migrations of Old World mammals across a Bering Sea land bridge added to the variety of mammals.

Stegomastodon was one of the few elephant-like mammals to migrate across the Bering Strait all the way to South America. Nearby is a full skeleton of *Merycodus*, a member of a family of fleet-footed grazers noted for their unusual horns. North America's pronghorn is the sole survivor of this ancient family.

Strobodon belonged to a group of dogs that evolved powerful jaws and massive teeth capable of crushing and eating bone. When the skeleton on display was discovered, it had between its ribs the bony remains of its last meal, a young *Merycodus*.

ICE AGE MAMMALS:
1.8 Million Years Ago to Today

Giant ground sloths stood nearly 6.1 m (20 ft.) tall and weighed several tons.

Massive glaciers advanced and retreated, and giant animals—most now extinct—roamed the continents during the Pleistocene. Entering *Ice Age Mammals,* you can almost feel the chill in the air.

Glaciers lowered the sea level as much as 91.4 m (300 ft.), enabling land connections between continents and islands to form. One such land link still stands between North and South America. Mammals on both sides of this bridge crossed over in what is known as the Great American Interchange.

A mix of dramatic fossils and modern specimens features some of the South American mammals that migrated to North America. Rivaling elephants in size, two giant ground sloths dominate the display. Species belonging to three different groups of giant sloths crossed into North America during the Pleistocene. None survived into the present. An ancient armored glyptodont dwarfs its contemporary relative, an armadillo. Armadillos still thrive in North America today, but glyptodonts died out 23,000 years ago.

FOSSILS AND TAR PITS

Tar pits are important sources of Pleistocene fossils. Both predator and prey were trapped in the sticky pools, where tar preserved their remains by sealing out air and moisture.

Scientists excavated more than 200 different animal species from the Rancho La Brea tar pits, now in downtown Los Angeles. Several beautifully preserved skeletons are featured in this display. A ground sloth faces off with a saber-toothed cat, while two dire wolves prowl nearby. They were larger but slower than our modern wolves.

Megaloceros giganteus, an extinct Irish elk, had antlers that measured up to 3.7 m (12 ft.) across and weighed up to 45 kg (100 lb.).

MASTODONS AND MAMMOTHS

Talk of the Ice Ages conjures up images of mammoths and mastodons, ancient elephant-like members of the order Proboscidea. You can view skeletons of both these majestic mammals here. Uniquely adapted to life in harsh, cold climates, woolly mammoths had layered coats with outer hairs more than 50.8 cm (20 in.) long. The mastodons occupied warmer woodland areas.

What happened to these Ice Age giants? It's possible that rapid changes in climate and vegetation hastened their demise, but it is more likely that they were driven to extinction by hunting and other changes caused by the arrival of another mammal: *Homo sapiens.*

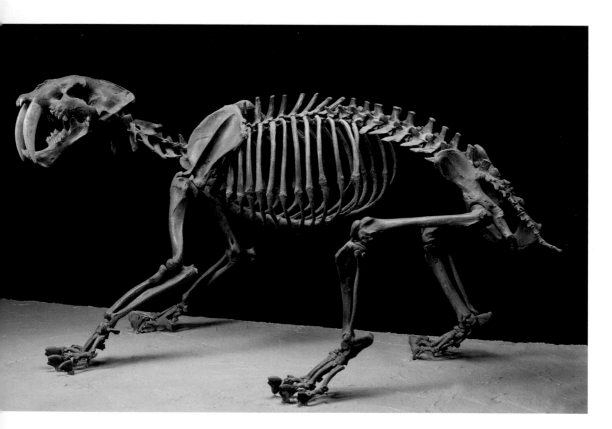

A saber-toothed cat bares its fangs. Huge canines helped these predators kill prey many times their size.

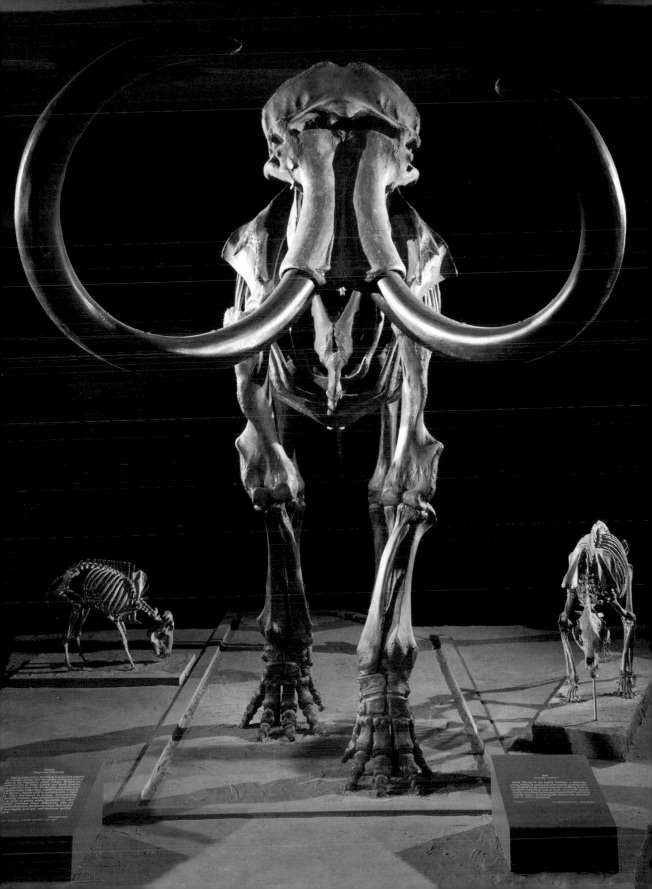

OTHER THINGS TO DO

Just beyond the Museum's exhibit halls, innovative learning experiences and great entertainment choices await.

Want to see hundreds of live butterflies from around the world as they flutter from flower to flower? Visit the Washington area's year-round indoor butterfly habitat. Interested in experiencing the thrill of a 3-D IMAX® film? The Museum offers a variety of films shown throughout the day, and on Friday and Saturday evenings as well. No matter what your learning style, you'll find experiences that educate, inspire, and entertain at the National Museum of Natural History.

Touchable artifacts and specimens allow visitors to the Naturalist Center in Leesburg, Virginia, to explore the collection at their own pace.

JOHNSON IMAX® THEATER

Enhance your visit to the Museum with an immersive adventure in the Johnson IMAX Theater.

Take a journey through time or voyage to exotic locations with our 3D experiences. Perhaps you'll explore the many depths of the ocean for close encounters with some of the world's most exotic marine animals. Or you might travel back to the age of the dinosaurs, coming face-to-face with some of Earth's largest creatures ever. Our six-story IMAX screen, 3D technology, and state-of-the-art sound system create the ultimate film event—making you feel like a participant in these fantastic journeys.

The IMAX films offered at the Museum tell stories that revolve around our collections and exhibits, from oceans to dinosaurs to human evolution. The Johnson IMAX Theater operates daily during regular Museum hours. Schedules, feature information, and tickets are available at 202 633-6071 and www.si.edu/imax.

DISCOVERY ROOM

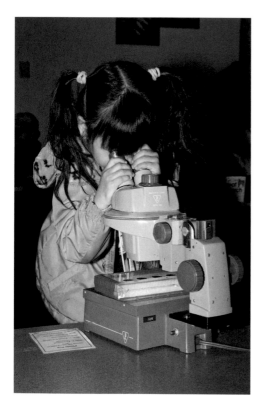

For thirty years, the Discovery Room has helped all our visitors, but especially our youngest, get excited about science. There you find touchable objects, kid-focused displays, and enthusiastic helpers in a welcoming and accessible space. Toddlers and youngsters often squeal, "Look, Mom. It's real!" as they are invited to handle specimens and artifacts from around the world.

Grab a pair of binoculars and search the biodiversity wall for animals found in the Washington area: coyotes, foxes, eagles, and cardinals perched on the far wall. Peer through microscopes and magnifiers to get a closer look at the wonders of the world. Open a drawer in the collection cabinets to find birds, bones, snake skins, and a question to answer from one of our scientists.

Microscopes in the Discovery Room open up a world of wonder for all ages.

Please consult the Museum website at www.nmnh.si.edu for Discovery Room hours, programming, and other recommendations. All children must be accompanied by an adult companion.

NATURALIST CENTER

Imagine a place where you can investigate your choice of over 36,000 natural history and anthropological objects. Picture a room where microscopes, measuring tools, and professional journals are right at your fingertips. That's what you'll find at the Museum's Naturalist Center.

Located just 56 km (35 mi.) from the Museum, in Leesburg, Virginia, the Naturalist Center is a unique, hands-on educational center designed for amateur naturalists ten years of age or older. Students, teachers, artists, and collectors will find everything they need to investigate nature's details up close. For younger children, a small Family Learning Center offers a range of special hands-on activities.

Although the Naturalist Center's collection contains specimens from around the world, its main emphasis is the natural and cultural history of the Mid-Atlantic region. You can explore tools used by Paleo-Indians, study fossils to imagine how the region looked millions of years ago, or identify an insect or plant found in your own backyard.

Interested in developing your own natural history collection? The Center's noncirculating reference library of over 6,000 volumes can help. Or attend one of the Center's Identidays, where Museum scientists are on hand to identify specimens and offer advice on preserving and organizing your collection. Teachers interested in designing hands-on, school group activities can participate in the Center's IWONDER professional development program.

If scientific illustration is your passion, the center provides collections and a variety of tools for drawing. At regularly scheduled Draw-Ins, you can talk to professional scientific illustrators about drawing techniques or careers.

Whether you choose to draw, identify, examine, or research, you can do it at your own pace. You're the investigator at the Naturalist Center, where learning about the world around you is as natural as asking questions.

Sizing up horse skeletons is just one of the many things you can do at the Naturalist Center.

PHOTO CREDITS

i, Chip Clark (2003-17668); **ii**, Chip Clark (99-27486); **v**, Chip Clark (2007-6936); **vi/vii**, Feodor Pitcairn Productions 2008; **viii/1**, Chip Clark (90-12712); **2**, Chip Clark (2003-8538); **3**, Chip Clark (95-40559); **4/5**, Chip Clark (99-27448); **6**, *t* Carl C. Hansen, *rb* Chip Clark (47227); **7**, Chip Clark (2003-16016); **8**, Chip Clark (2003-16020); **9**, Photo Archives (MAH 21935); **10**, *t* Photo Archives (MAH 18052), *b* Photo Archives (MNH 24881); **11**, United States National Herbarium; **12**, Chip Clark (2006-11698); **13**, National Portrait Gallery; **14**, Chip Clark (77-2087.01); **15**, *tl* Chip Clark (56881), *tr* Carlton Allen, *br* Chip Clark (53373); **16**, Chip Clark (955646); **17**, Chip Clark (37413); **18**, Chip Clark (2003-16045); **19**, Youngkye Yang (2003-15282); **20**, Encyclopedia of Life; **21**, Chip Clark (60827); **22**, *tl* Chip Clark (54637), *tr* Chip Clark (63229); **23**, Chip Clark (2008-12706); **24/25**, Sandra Raredon/SI; **26**, *t* John Steiner (2002-7841), *b* Donald E. Hurlbert (2007-9612); **27**, Carl C. Hansen (2002-806); **28**, *l* John Steiner (2002-13878), *r* James Di Loreto (2008-3401); **29**, Donald E. Hurlbert (2003-10584); **30/31**, James Forte; **31**, Donald E. Hurlbert (2003-10584); **32**, Chip Clark (2008-12708); **33**, John Steiner (2008-12848); **34**, Chip Clark (2007-7442); **35**, Exhibits/NMNH; **36**, John Steiner (2008-12662); **37**, Chip Clark (2008-12711); **38**, *t* Chip Clark, *b* James Di Loreto (2008-6446); **39**, Michael Lang; **40**, Chip Clark; **41**, Marilyn Aber; **42**, James Di Loreto (2007-12592); **43**, Chip Clark; **44**, Donald E. Hurlbert; **45**, James Di Loreto; **46**, John Steiner; **47**, *t* Carl C. Hansen (2003-4896), *c* Carl C. Hansen (2003-13953); **48**, Carl C. Hansen (2003-4899); **49**, Carl C. Hansen (2003-4876); **50**, *t* Carl C. Hansen (2003-4902), *b* Carl C. Hansen (2003-13966); **51**, Carl C. Hansen (2003-13943); **52**, *t* Carl C. Hansen (2003-48994), *b* Carl C. Hansen (2003-4888); **53**, Carl C. Hansen (2003-4877); **54**, *t* Carl C. Hansen (2003-13968), *b* Carl C. Hansen (2003-4878); **55**, *t* Carl C. Hansen (2003-13965), *b* Carl C. Hansen (2003-13980); **56/57**, Carl C. Hansen (2003-4883); **57**, Carl C. Hansen (2003-13981); **58**, Carl C. Hansen (2003-13972); **59**, Carl C. Hansen (2003-13958); **60**, Chip Clark (33730); **61**, Chip Clark (97-35339); **62**, *t* Chip Clark (55782), *b* Chip Clark (49069); **63**, James Di Loreto (2003-11575); **64**, Chip Clark (94-3182); **65**, Chip Clark (8106); **66**, *r* Harold Dorwin, *l* Francis Schroeder; **67**, Chip Clark (96-30254); **68**, *tr* Chip Clark (2003-16039), *cl* Chip Clark (2003-16041); **69**, *t* Chip Clark (2003-16038), *b* Chip Clark (2003-16043); **70**, *t* Chip Clark (21561), *b* Chip Clark (87-7494); **71**, Donald E. Hurlbert; **72**, Donald E. Hurlbert (2003-10606); **72/73**, Donald E. Hurlbert (2008-4693); **74/75**, James Di Loreto (2003-13872); **75**, Carl C. Hansen (95-781); **76**, John Gurche; **77**, John Gurche; **78**, David Brill/National Museums of Kenya; **79**, Human Origins Program/NMNH; **80**, *l* Human Origins Program/NMNH, *r* John Gurche; **81**, *t* Human Origins Program/NMNH,

b Human Origins Program/NMNH; **82**, *t* John Gurche, *bl* John Gurche, *br* John Gurche; **83**, *t* John Gurche, *b* Chip Clark (2008-13464); **84**, *t* James Di Loreto (2008-6237), *b* Donald E. Hurlbert (2008-3245); **85**, Chip Clark; **86**, Chip Clark (99-27461); **87**, *t* Diane Nordeck (99-21003), *b* Diane Nordeck (99-21027); **88**, *t* Tavy D. Aherne, *bl* Diane Nordeck (99-21064), *br* Carl C. Hansen (99-788); **89**, *l* Carl C. Hansen (99-778), *r* Maude Southwell Wahlman; **90**, Donald E. Hurlbert (2000-9836.11); **91**, *t* Donald E. Hurlbert/ James Di Loreto (2000-9833), *b* John Steiner (99-21387); **92**, John Steiner (2001-13933); **93**, Chip Clark (21539); **94**, Chip Clark (2003-4491); **95**, *t* Chip Clark (87-4485), *b* Chip Clark (87-4478); **96**, Chip Clark (51282); **97**, James Di Loreto (2003-12496); **98/99**, Chip Clark (2002-16207); **100**, *t* Chip Clark (96-3000), *b* Chip Clark (95-40466); **101**, Chip Clark (97-35264); **102**, Chip Clark (97-36277); **103**, *t* Chip Clark (2003-17665), *b* Chip Clark (2003-17666); **104**, Chip Clark (2002-32623); **105**, *t* Chip Clark (96-30249), *br* Chip Clark (95-40213), *bl* Chip Clark (95-40333); **106**, Chip Clark (2002-32620); **107**, Chip Clark (2003-16027); **108**, *l* Chip Clark (92-9450), *r* Chip Clark (MSA 165); **109**, *t* Chip Clark (95-40303), *c* Chip Clark, *b* Chip Clark (95-40594); **110**, Chip Clark (2003-16034); **111**, *t* Chip Clark (95-40167), *b* Chip Clark (12033); **112**, *t* Chip Clark (97-35216), *b* Chip Clark (97-36237); **113**, *t* Chip Clark (97-36234), *b* Chip Clark (2002-21840); **114**, *t* Chip Clark (2002-21826), *b* Chip Clark (60786); **115**, *t* Chip Clark (2002-11835), *b* Chip Clark (2002-11854); **116**, Chip Clark (2002-16207); **117**, Chip Clark (2002-16150); **118**, Chip Clark (2003-16053); **119**, Laurie Penland/Dane Penland; **120**, Chip Clark (2002-21653); **121**, Donald E. Hurlbert (12972); **122**, Chip Clark (97-36367); **123**, *t* Laurie Penland/Dane Penland, *bl* Chip Clark (97-35223), *br* Chip Clark (2003-16032); **124**, Chip Clark (97-36322); **125**, Chip Clark (97-36223); **126**, *t* Laurie Penland/Dane Penland, *b* Chip Clark (2003-16051); **127**, *t* Chip Clark (97-36209), *b* Laurie Penland/Dane Penland; **128/129**, Carl C. Hansen (2002-8071); **130/131**, Chip Clark (99-27383); **132**, *r* John Steiner (2003-15016), *l* Chip Clark (86-2411); **133**, John Steiner (2003-15032); **134**, John Steiner (2003-12106); **135**, Chip Clark (45216); **136**, John Steiner (2003-15012); **137**, John Steiner; **138**, Chip Clark (63282); **139**, John Steiner (2003-15047); **140**, John Steiner (2003-14998); **141**, Chip Clark (2008-2211); **142**, Chip Clark (87-10509); **143**, Chip Clark (073); **144**, *t* John Steiner (2003-15059), *b* Chip Clark (34315); **145**, Chip Clark (2006-24972); **146**, *t* James Di Loreto (2003-10382), *b* Chip Clark (2001-69); **147**, John Steiner (2003-15034); **148**, Chip Clark (85-75); **149**, Chip Clark (069); **150**, Chip Clark (0057); **151**, Chip Clark (0028); **152**, Chip Clark; **153**, Chip Clark (054); **154/155**, Chip Clark (2004-37378); **156**, James Di Loreto (2008-3344); **157**, James Di Loreto (2003-12495)

GENERAL INFORMATION

HOURS: Open daily (except December 25) from 10:00 A.M. to 5:30 P.M. Extended hours are determined annually and posted in the Museum.

MUSEUM WEBSITE: Check the Museum's home page, www.mnh.si.edu, for general information and special exhibit and event schedules.

INFORMATION SERVICES: The visitor information desks, located at the Constitution Avenue entrance and in the Rotunda, are staffed by volunteers daily from 10 A.M. to 4 P.M. For phone information, call (202) 633-1000 (voice) or (202) 357-1729 (non-voice TTY), Mondays through Fridays, 9 A.M. to 5:15 P.M., and Saturdays, 9 A.M. to 4 P.M. For online information, go to www.si.edu or www.mnh.si.edu.

TOURS: Walk-in highlight tours are given Tuesdays through Fridays at 10:30 A.M., Tuesdays through Thursdays at 1:30 P.M., and some Saturdays at 10:30 A.M. (except on some holidays). Meet in the Rotunda, Mall entrance.

LIVE DEMONSTRATIONS: Tarantula feedings, *O. Orkin Insect Zoo:* Tuesdays through Fridays, 10:30 A.M., 11:30 A.M., and 1:30 P.M.; Saturdays and Sundays, 11:30 A.M., 12:30 P.M., and 1:30 P.M.

LIVE BUTTERFLY PAVILION: Open daily 10:15 A.M. to 5 P.M. Tickets required; available online, by phone at (202) 633-4629 or (877) 932-4629, and at the Pavilion box office. Tuesdays free, but timed-entry ticket required. For more information or to purchase tickets, go to www.butterflies.si.edu/tickets.

PARKING: The Smithsonian does not have public parking facilities, but there are commercial parking lots within walking distance. Limited, three-hour free street parking is available on Jefferson and Madison Drives. Metered parking is available nearby. Handicapped spaces are located along the Mall for those vehicles bearing the appropriate license or state authorization.

PUBLIC TRANSPORTATION: The Museum can be reached by Metrorail, Washington's subway system, from the Smithsonian and Federal Triangle stations on the Orange/Blue lines. For more information, call Metrorail at (202) 637-7000 (voice) or (202) 638-3780 (non-voice TTY). Their website, www.wmata.com, shows bus and Metro routes and fares.

GROUP SALES/SERVICES: Group packages offer discounts on films for groups of 10 or more. The Atrium Café discount is available for groups of 20 or more. Call (866) 868-7774 or e-mail GroupSales@si.edu.

PHOTOGRAPHY: Handheld and video cameras for personal use are permitted in the Museum; the use of tripods is discouraged. Please check with visitor information desk for exceptions.

PETS: Only certified assistance animals are permitted in the Museum.

DINING: The 600-seat Atrium Café, located on the ground floor, serves a variety of foods to suit all tastes. The Fossil Café, found in the Dinosaur Hall, offers snacks, drinks and light meals in an educational environment. Smithsonian Resident and Contributing members receive a 10 percent discount with a valid membership card.

MUSEUM SHOPS: The Museum Store: Kids (ground floor) specializes in educational items for children and adults. The Gallery Store (ground floor) features gifts from around the world and an excellent bookstore. The Museum's specialty shops include The Gem and Mineral Store (2nd floor) and The Mammals Store (1st floor). Also, there are small sales kiosks located on the ground floor and in the Ocean Hall. Smithsonian Resident and Contributing members receive a 10 percent discount with a valid membership card.

SAMUEL C. JOHNSON IMAX® THEATER: For phone sales, call (202) 633-IMAX (4629) or (877) WDC-IMAX (932-4629). For groups, call (866) 868-7774. For online information and sales, go to www.si.edu/imax/.

DISCOVERY ROOM: Please consult the Museum website at www.mnh.si.edu for Discovery Room hours, programming, and other recommendations. All children visiting the Discovery Room must be accompanied by an adult.

VISITORS WITH DISABILITIES: The wheelchair accessible entrance is 10th Street and Constitution Avenue. Accessibility information is available at www.si.edu/visit/visitors_with_disabilities.htm, by telephone at (202) 633-2921 (voice) or (202) 633-4353 (non-voice TTY), or by email at access@si.edu. Baird Auditorium offers loop amplification in the center front rows. Rear window captioning, audio description, and assistive listening devices are available at the IMAX® Theater. Inquire at the box office. Tours are available for visitors who are blind, have low vision, or have learning differences. These tours must be requested in advance by calling (202) 633-1077 or emailing ramsdelll@si.edu. Wheelchairs are available on a first-come, first-served basis, free of charge. Many exhibitions include tactile experiences. Exhibit videos are captioned.

NATURALIST CENTER: Located in Loudoun County, Virginia, this hands-on, natural history collection and reference center is free of charge for visitors ages 10 and up. Tuesdays through Saturdays, 10:30 A.M. to 4 P.M. Closed Sundays, Mondays, and all federal holidays. 741 Miller Drive SE, Leesburg, Virginia 20175 (45-minute drive from D.C.). For services, special programs, and directions, call (703) 779-9712 or (800) 729-7725.

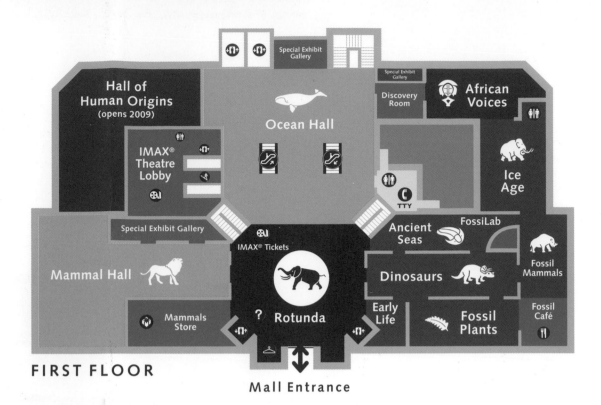

FIRST FLOOR

Hall of Human Origins (opens 2009)

Ocean Hall

Special Exhibit Gallery

Discovery Room

African Voices

Ice Age

IMAX® Theatre Lobby

Special Exhibit Gallery

Mammal Hall

Mammals Store

IMAX® Tickets

? Rotunda

Ancient Seas

FossiLab

Fossil Mammals

Dinosaurs

Early Life

Fossil Plants

Fossil Café

TTY

Mall Entrance

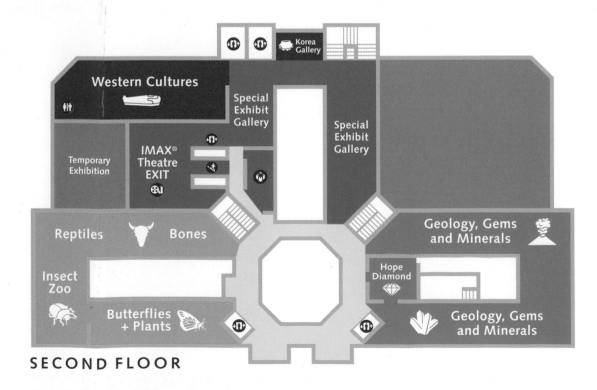

SECOND FLOOR

Korea Gallery

Western Cultures

Special Exhibit Gallery

Special Exhibit Gallery

Temporary Exhibition

IMAX® Theatre EXIT

Reptiles

Bones

Geology, Gems and Minerals

Insect Zoo

Hope Diamond

Butterflies + Plants

Geology, Gems and Minerals